Мы спускались, а перегрузки действовали на нас все сильнее. Включился двигатель мягкой посадки, аппарат ударился о землю и упал набок. Люк открылся, мы увидели небо и заглядывающие в иллюминатор человеческие лица. Я утратил координацию. Мое тело, руки и ноги словно налились свинцом. Милая Земля, как тяжела ты на моих плечах!

Валерий Рюмин
СССР

As we descended farther, the pressure on our bodies mounted. The engine of the soft landing started up, and we hit the ground and turned to one side. The hatch opened and we saw the sky and human faces peeking in. I had lost coordination. My body, arms, and legs were terribly heavy. Dear Earth, how heavy you are, sitting on my shoulders.

Valeri Ryumin
USSR

При движении космического корабля в плотных слоях атмосферы чувствуешь себя так же, как при быстрой езде на телеге по булыжной мостовой.

Лев Дёмин
СССР

When the spacecraft is moving across the denser layers of the atmosphere, you feel as if you are in a cart traveling rapidly over cobblestones.

Lev Demin
USSR

123   The Rocky Mountains of British Columbia, Canada

As we encounter the thin upper reaches of the atmosphere, the orbiter begins a gradual transition from ballistic orbital object to aerodynamic flying machine. The first clue comes as the wings bite into the air producing an acceleration that causes you to sink into the seat cushion. You hear wind noise around the windshield which slowly increases; then you feel an occasional tremor from turbulence. Gradually the impulses from the control jets become unnoticeable and the familiar feeling of aerodynamic flight is complete.

Gordon Fullerton
USA

Мы вошли в атмосферу. Растет перегрузка и вдавливает в кресло. На стеклах иллюминаторов появились оранжевые нити плазмы. Летим в огненном шаре.

Анатолий Березовой
СССР

We have entered the atmosphere. The pressure is growing and pushes you into your seat. On the glass ports have appeared orange filaments of plasma. We are flying in a fiery sphere.

Anatoli Berezovoy
USSR

Прошел год и мне стало казаться, что эти 4200 часов 35 минут и 36 секунд были совместно прожиты в космосе какими-то двумя другими людьми.

Валерий Рюмин
СССР

A year has passed and it seems to me that these 4,200 hours, 35 minutes, and 36 seconds were lived in space by two entirely different men.

Valeri Ryumin
USSR

120-122   The Horn of Africa

Близилась к концу наша экспедиция, до возвращения на Землю оставались считанные часы, и тут вдруг пришла мысль: «Неужели больше никогда не будет всего этого – пьянящего чувства невесомости, плывущей за иллюминатором Земли, похожей на большой, расцвеченный солнечными красками школьный глобус, фантастических фиолетовых всполохов гроз и светящейся паутины больших городов с их окрестностями на фоне иссиня-черной ночной планеты? «Хотелось запомнить, отложить в самый надежный уголок памяти все, что возможно увидеть, почувствовать лишь в космосе, чтобы пронести это с собой на всю оставшуюся жизнь.

Светлана Савицкая
СССР

Our mission was coming to an end and only a few hours remained before my return to Earth. All of a sudden I thought, Could it be that I'll never have the chance to have any of this again: the drunken feeling of weightlessness and the Earth swimming outside the porthole. Never again would I see the Earth looking like a gigantic, multihued, sunlight-colored classroom globe, nor the fantastic violet flashes of lightning, nor the spider's webs of large cities and their suburbs shimmering against the blue-black night of the planet. I wanted to remember, to store in the most reliable part of my memory, everything I could only ever see or feel in space so that I could keep it with me for the rest of my life.

Svetlana Savitskaya
USSR

The moments of parting are sad. Symbolically, by Russian tradition, we all sat down for a moment, having hooked our feet on the chairs. For a second we paused in the round frame of the transfer hatch, the two time-servers, and the three cosmonauts who were leaving. Then Sasha Laveykin, Muhammad Faris, and I swam out of the station one by one. Yuri Romanenko snapped shut the lock to the *Mir* and unexpectedly called out, "Now if you ask to come back, we won't let you." However, the joke did not cheer us up.

Aleksandr Viktorenko
USSR

Грустные минуты расставания. Символически, по русскому обычаю, все присели, зацепившись ногами за кресла. На секунду в круглой рамке переходного люка застыли вместе двое старожилов и трое отъезжающих. Потом я, Саша Лавейкин и Мухаммед Фарис один за другим медленно выплыли со станции. Юрий Романенко защелкнул люк «Мира» и неожиданно крикнул: «Сейчас ведь обратно попроситесь, а мы вас не пустим!» Но веселее от шутки нам не стало.

Александр Викторенко
СССР

Ночь. Прошли Атлантику. Вышли на Европу. Внизу множество серебряных огоньков. Они образуют дивный сверкающий ковер. Наконец, на фоне этого ковра появляется мерцающая звезда с лучами шоссе. Москва.

Александр Лавейкин
CCCP

Night, and we had crossed the Atlantic. We came upon Europe. Below were a multitude of silver sparks. They formed a glorious glittering carpet. Finally, a shimmering star with radiating highways appeared against a background of this carpet; I had seen Moscow.

Aleksandr Laveykin
USSR

На полу слева от бегущей дорожки —
иллюминатор. Я люблю бегать и
смотреть на Землю. Неважно, чья она
— главное, что она есть.

Олег Атьков
СССР

On the floor to the left of the running
track is a porthole. I love running and
looking at the Earth. It isn't important
whose she is, just that she is.

Oleg At'kov
USSR

117   Ancient meteor crater, Quebec, Canada

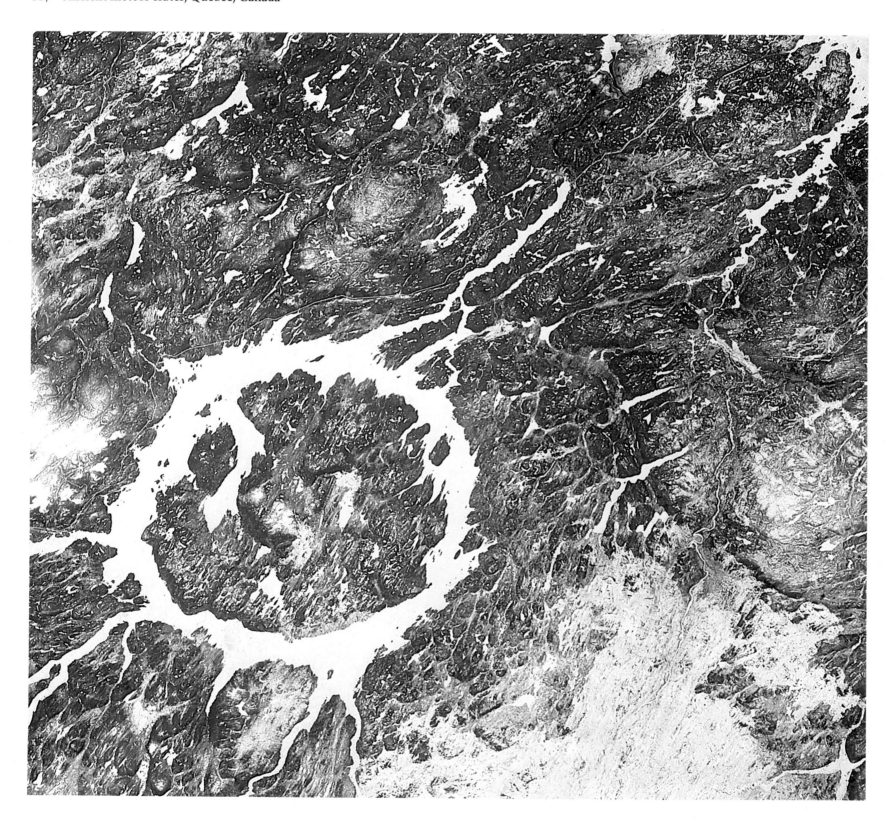

116   The Marshall Islands in the Pacific Ocean

Сегодня у нас была телевизионная сессия. Мы поздравляли Толину дочку Таню с ее восьмым днем рождения. Мы сделали торт из хлеба. Вместо свечек, мы использовали ручки, и вместо пламени – серебрянную бумагу. И у нас еще были электрические свечки – четыри фонаря, которые выглядили, как будто их было восемь когда мы поставили за ними зеркало. Мы развесили цветные воздушные шарики, ездили верхом на пылесосе и на воздушном шарике. В общем, развлекали эту маленькую девочку по телевизору.

Валентин Лебедев
СССР

Today we had a television session. We wished Tolya's daughter Tanya a happy eighth birthday. We made a cake out of bread, and instead of candles we used pens, and for flames we used foil. We also had electric candles – four flashlights, which looked like eight when we put a mirror behind them. We hung up colored balloons, rode around on the vacuum cleaner and on a balloon. All in all, we entertained the little girl via television.

Valentin Lebedev
USSR

Однажды в полете меня спросили с Земли, что я вижу. «Да что вижу?» ответил я, «Полмира справа, полмира слева, весь мир как на ладони! Земля! Ох, мала!»

Виталий Севастьянов
СССР

Once during the mission I was asked by ground control what I could see. "What do I see?" I replied. "Half a world to the left, half a world to the right, I can see it all. The Earth is so small."

Vitali Sevastyanov
USSR

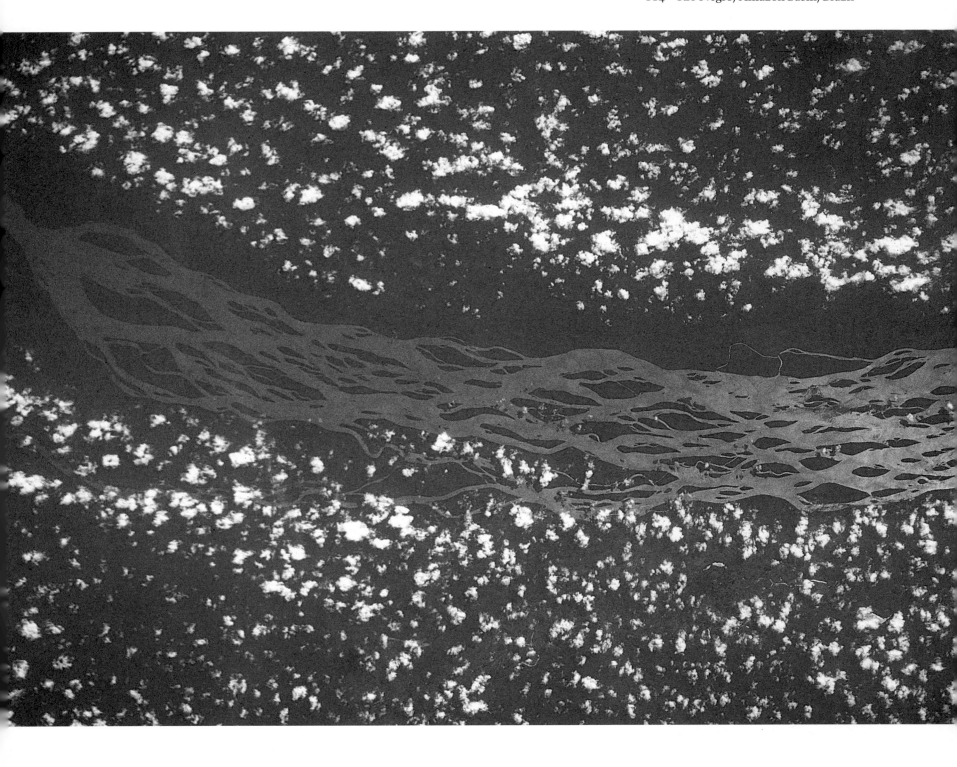

В разгар праздничного, в честь нашего прибытия, обеда, я достал сюрприз для хозяев станции – две маленькие связки укропа. Во время предполетного завтрака я, тайком от медиков, завернул их в салфетку и спрятал в карман. Хоть я и понимал, какую драгоценность везу для Валерия и Леонида, не видевших зелени более двух месяцев, такого восторга все же не ожидал.

«Укроп! Чудо какое! Но почему так мало?!»

Валерий Кубасов
СССР

In the midst of the dinner held in honor of our arrival, I got out a surprise for our hosts. I had brought two thin bundles of fresh dill, and during the preflight breakfast, and without letting the medics know, I wrapped them in a paper napkin and put them in my pocket. Although I knew what a treasure I was bringing for Valeri and Leonid who had not seen anything green for more than two months, I had not expected their enthusiasm.

"Dill! What a miracle! But why so little?!"

Valeri Kubasov
USSR

Когда находилась свободная минута,
я писал письма жене, дочке,
родственникам, друзьям. Так было
легче переносить разлуку с Землей.

Анатолий Березовой
СССР

When I had a free minute I would write
letters, to my wife, my daughter,
relatives and friends. It was easier that
way to tolerate the separation from
the Earth.

Anatoli Berezovoy
USSR

Разговариваешь с женой и можешь
угадать по интонации и паузами
между словами то, что она хочет
сказать, но не произносит.

Валентин Лебедев
СССР

You talk with your wife and can guess
from her intonation and intervals
between words a lot more than what
she's actually saying.

Valentin Lebedev
USSR

Para ver si es posible obtener materiales ultrapuros en el espacio, donde no hay fuerza de gravedad, realizamos un experimento poniendo cuatro cristales de azúcar en un cristalizador que sumergí en una disolución de sacarosa. La cosa resultó bastante difícil. En la casa uno revuelve con una cucharadita y ya tiene el té dulce. Pero aquí hay que sacudir la cubeta hasta sudar la gota gorda. Así que la "vida dulce" en el cosmos no se consigue tan fácilmente.

Arnaldo Tamayo Méndez
Cuba

In order to see whether it is possible to get ultra-pure materials in weightlessness, we designed an experiment in which four crystals of sugar were put in a crystallizer that I immersed in sucrose solution. The task turned out to be quite difficult. At home you use a spoon and there you have it – sweet tea. In space you shake the flask until you're blue in the face. The sweet life is not easy to achieve in space.

Arnaldo Tamayo Méndez
Cuba

112   Kamchatka Peninsula, Siberia, USSR

Запахи леса и травы, теплый летний дождь и пушистый снег на поляне, близких и друзей после нескольких недель космического полета представить отчетливо уже не можешь и встречаешься с ними лишь во сне.

Петр Климук
СССР

After several weeks it became difficult to remember clearly the fragrance of grass and trees, or warm summer rain, or powdery snow in a glade, or the faces of friends and loved ones that you now see only in dreams.

Pyotr Klimuk
USSR

Летаешь по орбите, а снится обычно земное.

Владимир Ляхов
СССР

While out there in orbit, dreams were usually about Earth.

Vladimir Lyakhov
USSR

Над моим спальным мешком, рядом с фотографиями Гали и сына Лени прикрепил столь долгожданную фотографию крохотной Танюшки. Какая радость, что ребята прислали. Дочка родилась, а папа в космосе, и расцелует ее только через пять месяцев. Надо заставить себя не считать дни, а то с тоской не справишься. Расти, Танюшка! Твой папа мысленно всегда с тобой.

Леонид Кизим
СССР

Above my sleeping bag, next to the photographs of my wife Gayla and my son Lenya, is fixed the long-awaited photograph of my tiny Tanya. What joy the guys brought me when they sent up the photo. A daughter born while her father is in space; he won't be able to kiss her for five more months. I had to stop myself from counting the days; otherwise I would not have been able to bear it. Grow, little Tanya, I would think, your father is always with you in spirit.

Leonid Kizim
USSR

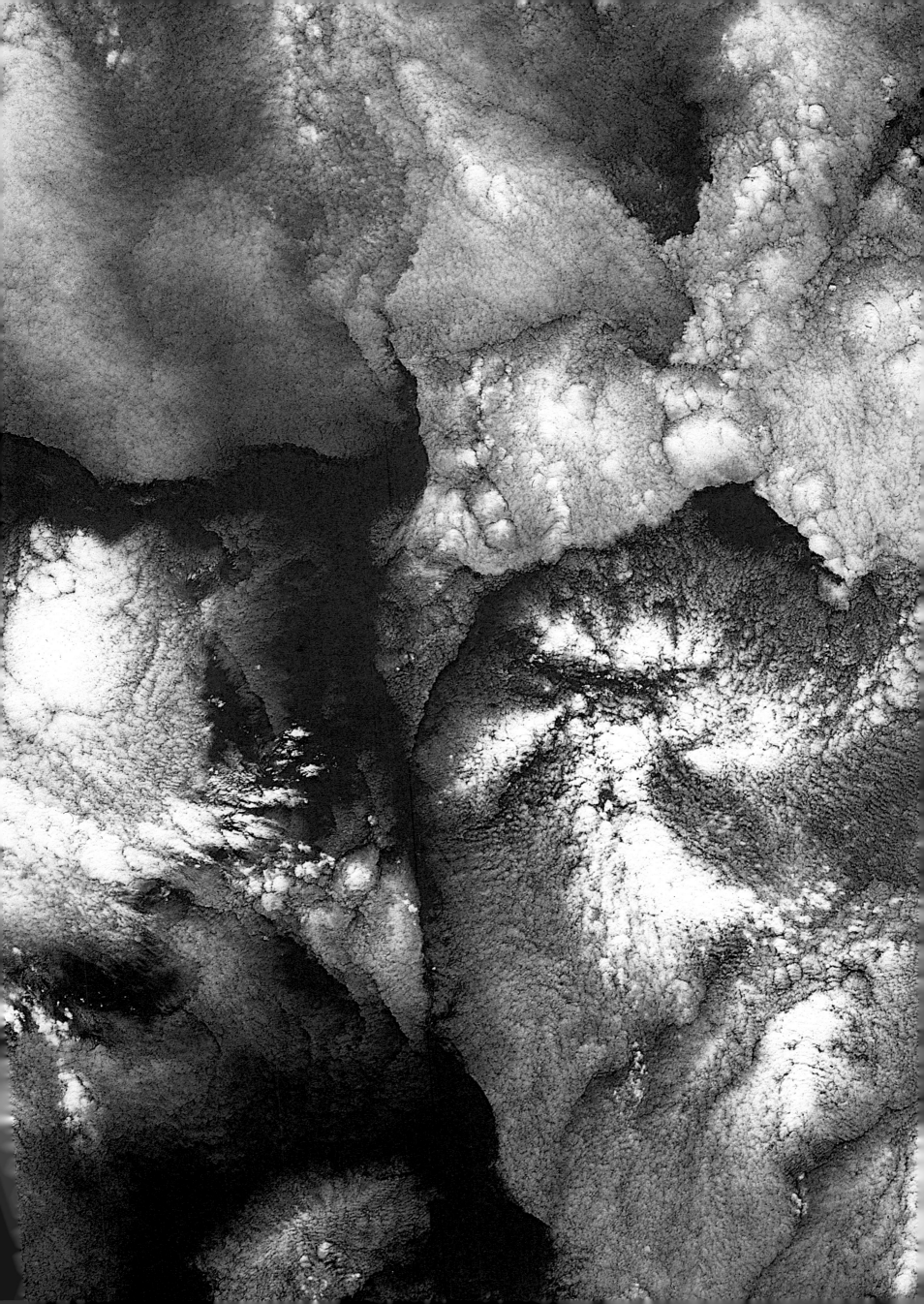

Проснувшись как-то утром, я решил выглянуть из окна, чтобы посмотреть где летим. Мы летели над Америкой, и вдруг я увидел снег, первый снег, который мы когда либо видели из космоса. Легкий и как порох, он смешивался с очертаниями земли, с прожилками рек. И я подумал: осень, снег, люди все заняты подготовкой к зиме. Некоторое время спустя, мы летели над Атлантическом океане, потом над Европой, и потом над Россией. Я никогда не был в Америке, но думаю, что приход осени и зимы такой же, как и в других местах, и процесс подготовки к ним тоже такой же. И мне пришло в голову что все мы дети нашей Земли. Неважно на какую страну смотреть. Мы все дети Земли и должны к ней относиться как к Матери.

Александр Александров
СССР

One morning I woke up and decided to look out the window, to see where we were. We were flying over America and suddenly I saw snow, the first snow we ever saw from orbit. Light and powdery, it blended with the contours of the land, with the veins of the rivers. I thought – autumn, snow – people are busy getting ready for winter. A few minutes later we were flying over the Atlantic, then Europe, and then Russia. I have never visited America, but I imagined that the arrival of autumn and winter is the same there as in other places, and the process of getting ready for them is the same. And then it struck me that we are all children of our Earth. It does not matter what country you look at. We are all Earth's children, and we should treat her as our Mother.

Aleksandr Aleksandrov
USSR

С орбиты мы видели все времена года; старт был весной, а все лето и осень, вплоть до начала зимы, мы летали. Сначала белизна снегов уступила место зелени лета, а затем покрылись золотом поля и леса, и снова все стало белым; только граница снегов перемещалась теперь в противоположном направлении.

Анатолий Березовой
СССР

From orbit we observed all the seasons of the year. The launch was in the spring, and we flew throughout the summer and fall and the start of winter. At first the whiteness gave way to the green of summer, and then gold covered the fields and forests, and then the whiteness again. Only now its border was moving in the opposite direction.

Anatoli Berezovoy
USSR

На орбитальную станцию мы привезли с Земли рыбок для исследований. И вот, прожив недели две, рыбки стали погибать. Как же нам было жаль их! Что мы только не делали, чтобы их спасти! А ведь на Земле нам огромное удовольствие доставляли и рыбная ловля, и охота.

Когда побудешь в одиночестве, в отрыве от всего земного, особенно остро воспринимаешь любое проявление жизни.

Виталий Жолобов
СССР

We brought some small fish to the orbital station for certain investigations. After two weeks they began to die. How sorry we felt for them! What we didn't do to try to save them. And yet on Earth, we have great pleasure fishing and hunting.

When you are alone and far from anything terrestrial, any appearance of life is especially welcomed.

Vitali Zholobov
USSR

 Lake Winnipeg, Canada

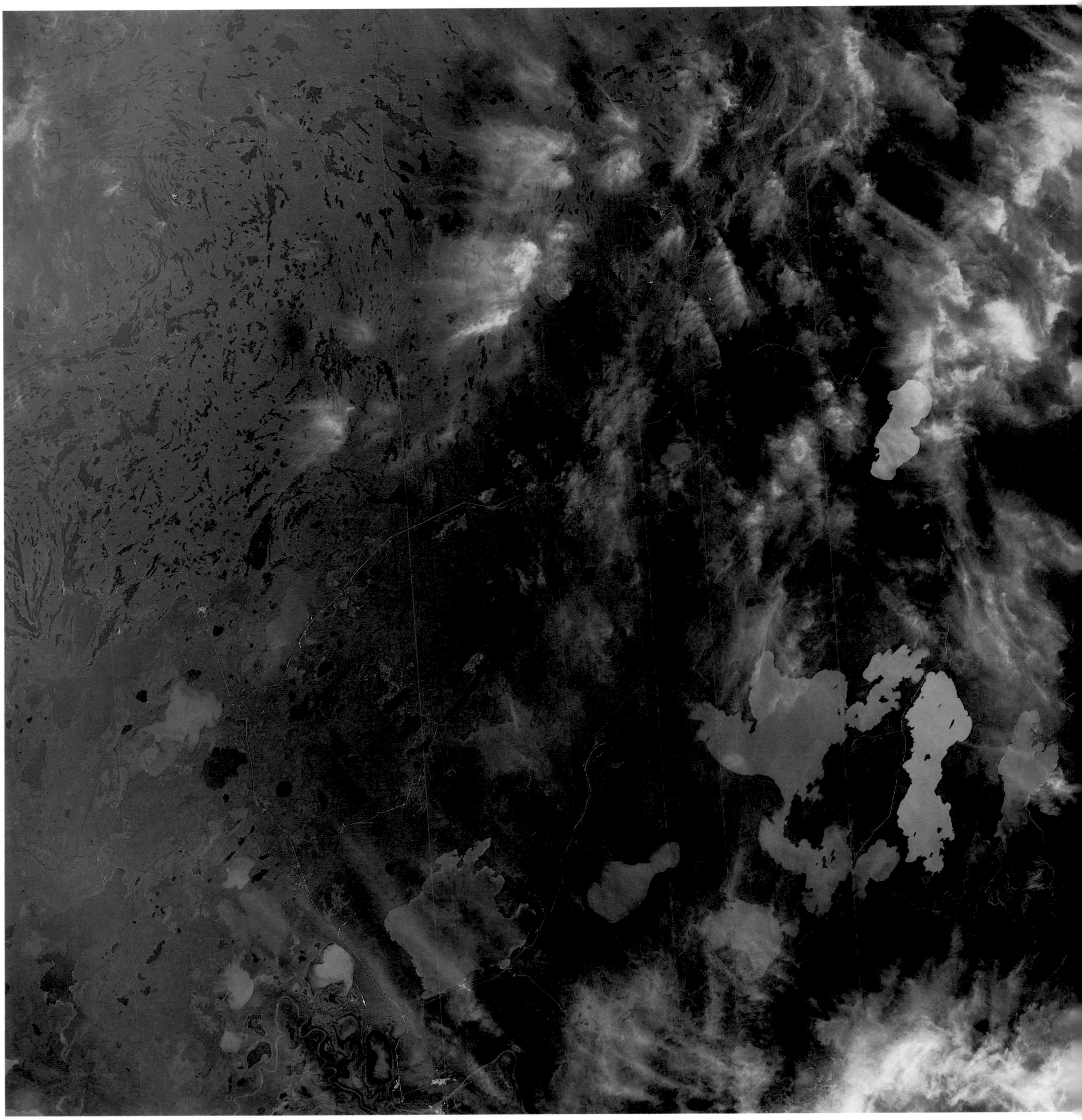

Невозможно представить, сколько удовольствия доставлял нам наш маленький огород – зеленый оазис земной жизни. Мы даже говорили: «Ну, пойдем, погуляем в рощу!»

Человеческому глазу, привыкшему к зеленому убору Земли, живые растения дарят спокойствие. Даже спится лучше рядом с этими тоненькими стебельками.

Георгий Гречко
СССР

It's impossible to imagine how much pleasure this green oasis of terrestrial life gave us. We even said, "Come on, let's go walk in the grove!"

For human eyes accustomed to the greenness of Earth, living plants give tranquillity. Even sleeping beside those thin stems was better.

Georgi Grechko
USSR

Первые отростки появились в нашем огороде. Мы посадили бурак, редиску и огурцы. Горох растет прямо из земли, с толстым стебелем и свертыми листьями. Пшеница просто растет верх, как зеленый луч света. Я люблю ладонями притрагиваться к отросткам. Они щекочут.

Валентин Лебедев
СССР

The first shoots came up in our garden. We planted beets, radishes, and cucumbers. The pea plant comes out of the soil with a fat stem and tightly wrapped leaves. The wheat just shoots up like a ray of green light. I like running my palm over the shoots. They tickle.

Valentin Lebedev
USSR

106   Plateau du Djado, Niger

Лебедев никогда не выращивал
растения, но на станции он бежал
каждое утро в наш «оазис», не успев
еще открыть глаза. Он там
выращивал горох и овес.

Анатолий Березовой
СССР

Lebedev had never before grown
plants, but on the station he used to
rush off to our "Oasis" installation
every morning almost before his eyes
were opened. He was growing peas and
oats there.

Anatoli Berezovoy
USSR

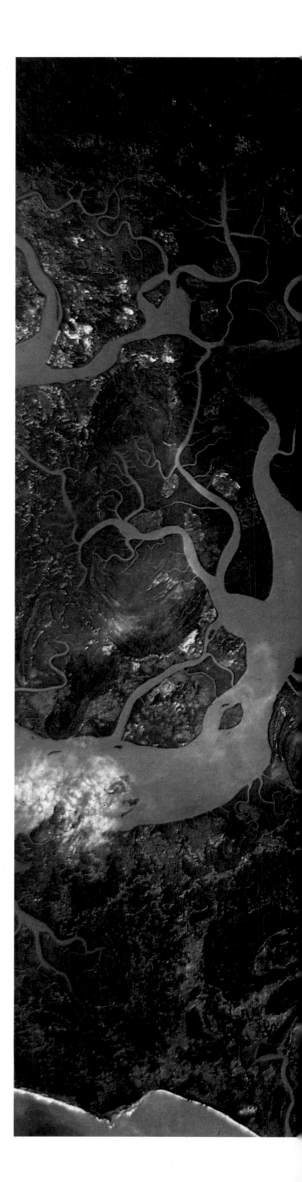

Как-то на иллюминаторе я увидел кристаллики льда. Эти кристаллы были совсем иные. Они были асимметричны. И вообще похожи на инвалидов из чудесного мира земных кристаллов. Я вспомнил нашу русскую зиму, натертые снегом щеки, пьянящий воздух – это не здешняя смесь газов.

Виталий Севастьянов
СССР

I once saw ice crystals on the porthole glass. They were alien and asymmetric, one might say like invalids from the miraculous world of terrestrial crystals. I remembered our Russian winters, snow-rubbed cheeks, and air that can be drunk in – not the mixture of gases to be had aboard a spacecraft.

Vitali Sevastyanov
USSR

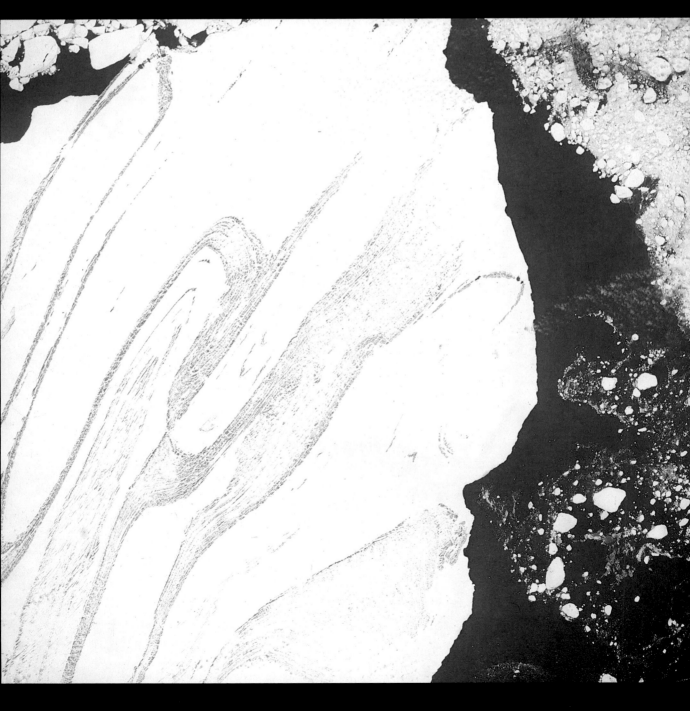

На станции «Салют-3» был пылесос под названием «Ракета». Когда мы забывали его выключить, он совершал стремительные полеты. Я решил, что это неплохое средство передвижения. Удобно устроившись верхом на пылесосе, я начал кататься по станции, а Попович, оправившись от изумления, схватил фотокамеру, чтобы сохранить для потомков новый рывок технического прогресса.

Юрий Артюхин
СССР

The *Salyut* station had a vacuum cleaner called the Rocket. When we forgot to switch it off it would try to fly away, so I decided it would be quite a good method of moving around. Straddled comfortably on the cleaner, I rode around the station. After recovering from his amazement, Popovich snatched up a camera in order to preserve for posterity this new technical advance.

Yuri Artyukhin
USSR

В течение всего дня я должен был наблюдать за приборами. Очень трудно было выкроить время, чтобы оторваться от них и посмотреть вниз, на Землю. Но когда я все же улучил момент и взглянул туда, я был поражен красотой увиденного многоцветного мерцающего мира.

Олег Макаров
СССР

All day long I had to watch the instruments. The most difficult part was to find time to lean back and just look down at the Earth. But when for a brief moment I did look down, I was stunned to see our colorfully shimmering world.

Oleg Makarov
USSR

102   Zagros Mountains and Qeshm Island, Iran

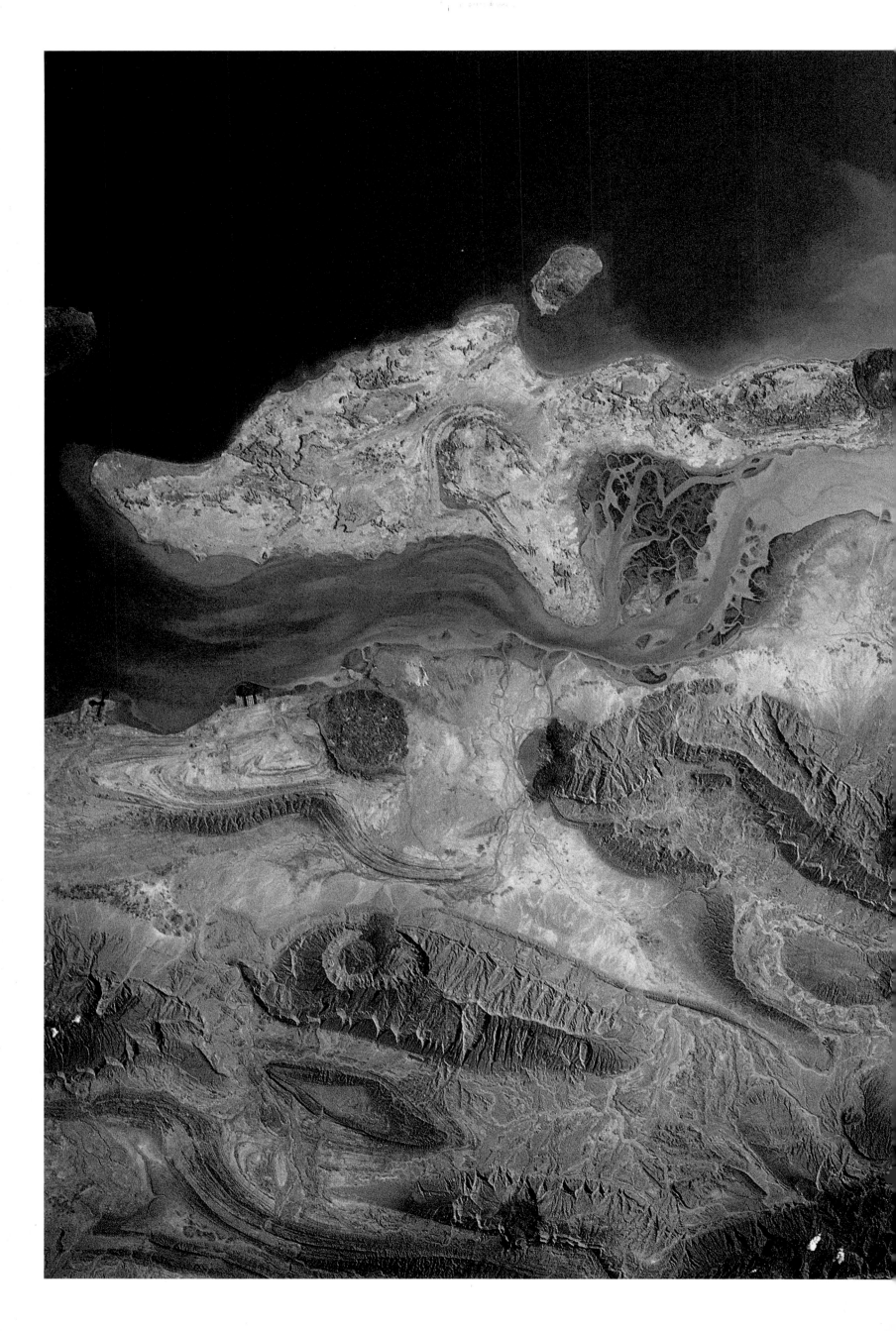

«Как погода?» хочет узнать Лебедев.

«Не балует», отвечает оператор.

«Моросит дождь. И холодно».

«А у нас тепло и сухо», смеется Валентин.

«Сегодня у вас тоже будет сыро», отвечает оператор, «сегодня банный день.»

Анатолий Березовой
СССР

"And how's the weather?" Lebedev wants to know.

"It's not spoiling us," answers the operator. "It's drizzling. And it's cold."

"We're warm and dry," Valentin laughs.

"But today you're going to get a little wet," responds the operator. "It's bath day."

Anatoli Berezovoy
USSR

Утром я сделал зарядку – проехал на велоэргометре от Южной Америки до Владивостока, благополучно преодолев Гималаи. Вечером прошел пешком с перебежками от Лос-Анджелеса до Лиссабона и даже не заметил шторма на Атлантическом океане.

Виталий Севастьянов
СССР

In the morning I did my exercises – a ride on the exercise bike from South America to Vladivostok, successfully overcoming the Himalayas. In the evening I took a stroll on the walking machine from Los Angeles to Lisbon, and I didn't even notice the storm in the Atlantic.

Vitali Sevastyanov
USSR

100   Tropical storm over the Indian Ocean

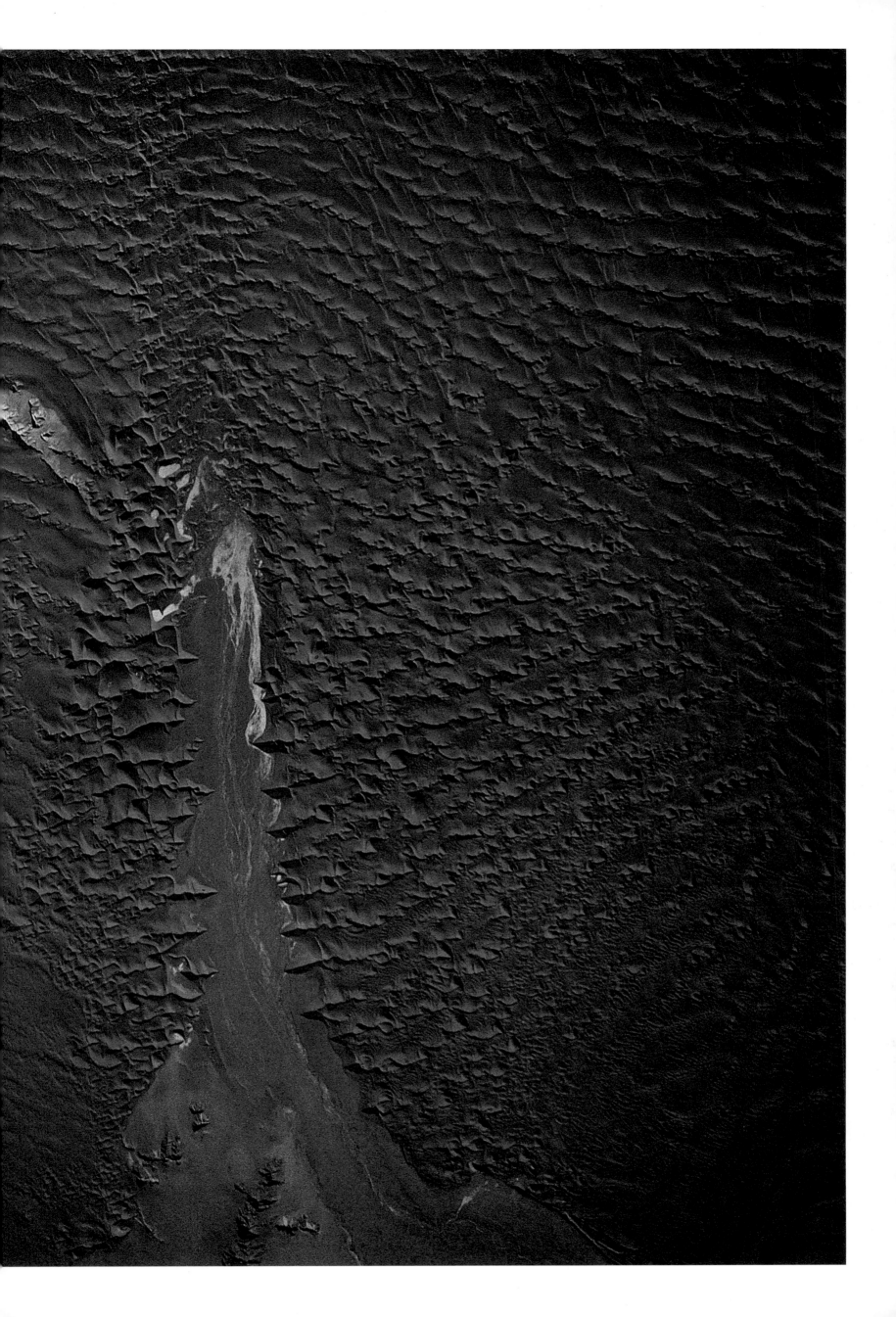

У нас были различные
магнитофонные записи: концерты,
легкая музыка. Но к концу полета
мы больше стали сдушать русские
народные песни. Имелись у нас и
записи природных звуков: грохот
грома, шум дождя, пение птиц. Мы
включали их чаще всего и никогда
не уставали от этих записей. Они как
будто возвращали нас на землю.

Анатолий Березовой
СССР

We had various kinds of taped
recordings: concerts, popular music.
But by the end of the flight what we
listened to most was Russian folk
songs. We also had recordings of
nature sounds: thunder, rain, the
singing of birds. We switched them on
most frequently of all, and we never
grew tired of them. It was as if they
returned us to Earth.

Anatoli Berezovoy
USSR

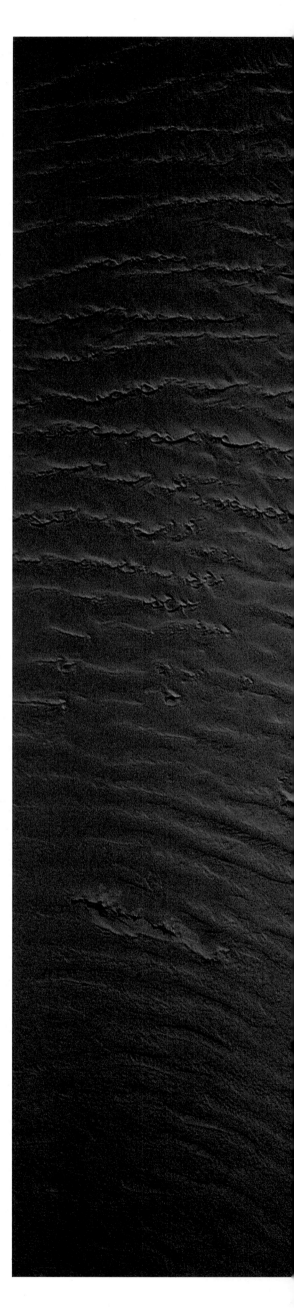

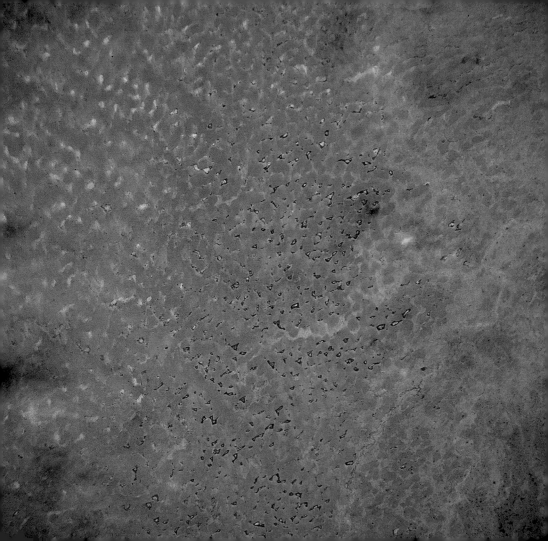

97 The Gulf of Suez

Мы нашли курагу, почувствовали себя ближе к земной жизни, но чем-то эта жизнь отличается от земной. Все земное вспоминается, как будто все это было давно и нас там уже нет, и неизвестно, когда вернемся.

Александр Александров
CCCP

We found dried apricots, and it's as if we're closer to Earth life. But there is something that separates the experience from the usual Earth way. Everything Earthly is remembered as if it were very long ago and we are no longer there and it is not known when we will be again.

Aleksandr Aleksandrov
USSR

Сегодня мы встретили первый грузовой корабль без пилота – «Прогресс-5». Он состыковался с нашей «задней дверью» и выгрузил массу вещей, в том числе аппарат, который одновременно производит кислород и всасывает углекислоту. Вместе с едой, водой, письмами и сувенирами, кто-то тайком положил книгу о подмосковней деревне, где я вырос. Нетерпеливо я перелистывал ее, в поисках фотографий ручьев, рек и озер. Она мне напомнила рыбную ловлю в детстве, вечера у костра, восходы и закаты, родной дом. Я был очень благодарен неизвестному другу, который тайком ее положил, чтобы скрасить наше пребывание здесь.

Валерий Рюмин
СССР

Today we welcomed our first delivery truck, a pilotless ship, *Progress 5*. It docked at our "back door" and brought us loads of things including a machine that produced oxygen while absorbing carbon dioxide. Along with the food, water, letters, and souvenirs, somebody had sneaked in a book on the Moscow countryside where I grew up. I eagerly leafed through it for pictures of streams, rivers, and lakes. It reminded me of my childhood outings to fish, and nights by the fire, sunrise, sundown, home. I felt thankful to the anonymous friend who had smuggled it in so as to enliven our existence in space.

Valeri Ryumin
USSR

96   The mouths of the Ganges in Bangladesh

95   The Brazilian coast in the state of Paraná

«Ни дня без открытия» таков был наш девиз в полете. Если это не получалось в экспериментах, мы открывали на ужин консервные банки.

Владимир Коваленок
СССР

"Not a day without a discovery" was our motto during the mission. If we were unable to make a discovery in our experiments, then we would discover what was for lunch.

Vladimir Kovalyonok
USSR

В первом полете с Валерием Рюминым я потерял часы. По предыдущему опыту я знал, что искать надо на решетке пылесборника. Открыл малые панели – ничего нет. А за ними – большие панели, завинченные шурупами. Подплыл с отверткой к той, где прикреплен мой спальный мешок. Отвязываю мешок, чтобы освободить панель, и вижу записку: «Чудак! Положи отвертку. Здесь никогда ничего не бывает. В. Коваленок».

Владимир Ляхов
СССР

During my flight with Valeri Ryumin I lost my watch. We knew from past experience that whenever we lost something, the first place to look is at the ventilator grille. I opened the small panels – nothing.

I floated, screwdriver in hand, up to the panel to which my sleeping bag was attached. I untied it in order to free the panel and saw a sign: "Put down your screwdriver, you nut, this place is guaranteed empty. Signed: V. Kovalyonok."

Vladimir Lyakhov
USSR

Станция, попавшая без людей в беду, встретила нас ледяным молчанием. В абсолютной тишине мы с Виктором Савиных обследовали отсеки. Луч ручного фонарика выхватывал различное снаряжение, аккуратно размещенное по бортам. Интерьер был в идеальном состоянии. На крошечном столе – традиционные хлеб-соль, оставленные Леонидом Кизимом, Володей Соловьевым и Олегом Атьковым, и письмо с просьбой беречь этот дом и пожеланиями удачи.

Владимир Джанибеков
СССР

The station had fallen into trouble without people aboard and met us now with an icy silence. In absolute silence Viktor Savinykh and I inspected the compartments. The beam of the flashlight picked out the various items of equipment, all in their correct places along the sides. The interior was in ideal condition. The traditional Russian welcoming gift of bread and salt had been left on the tiny table by Leonid Kizim, Volodya Solovyov, and Oleg At'kov, together with a letter asking those who came after them to look after this house and wishing them success.

Vladimir Dzhanibekov
USSR

Семь долгих дней мы пытались то на дневной стороне орбиты, то с ручными фонариками найти причину отказа, зажечь хоть одну лампочку.

И все-таки, проверив бортовые системы, мы нашли виновников аварии – аккумуляторы. Развернули солнечные батареи станции на Солнце, раскрутили толстенный жгут проводов и напрямую, минуя автоматику, соединили батареи с аккумуляторами. И, наконец, – да будет свет! Так что, если уж говорить, что важнее, то автомат – это, конечно, замечательно. Но последнее слово все-таки всегда остается за человеком.

Виктор Савиных
СССР

For seven long days, working both during the day-side parts of the orbit and by flashlight, we tried to find out what was causing the solar panels to fail. We wanted to get at least one bulb to light. Finally, we found the culprits when we checked all the storage batteries.

Two of them had gone out of commission. We undid the thickly plaited cabling and connected the solar panels directly so that they always faced the sun. The batteries began to recharge and finally there was light.

Automation is indeed a wonderful thing, but in the end humanity has the last word.

Viktor Savinykh
USSR

После третьей основной экспедиции станцию «Салют» законсервировали и пять месяцев с ней поддерживали только контрольную радиосвязь. И вдруг связь оборвалась. «Салют» молчал.

Мы с Джанибековым отправились в космос искать взбунтовавшийся «Салют-7» Двое суток мы, меняя орбиты, высматривали его с корабля. И вот из-за горизонта в лучах Солнца сверкнула наша рукотворная звезда.

Вплываем в «Салют», включаем свет – он, конечно, не зажигается. Полная тьма, могильный холод и зловещая, гнетущая, поистине космическая тишина.

Виктор Савиных
СССР

After the third major mission, *Salyut* 7 was mothballed, and for five months it was maintained by radio control. Then communications ceased and *Salyut* went silent.

Dzhanibekov and I were launched into space to find the *Salyut* 7, which had gone "on strike." After changing orbits, we searched for the station from the spacecraft for two days. At last, the man-made star rose above the horizon and flashed in the rays of the sun.

We floated into the station and turned on the lights, which did not, of course, light up. There was complete darkness, deadly cold, and an ill-boding, oppressive, and truly cosmic silence.

Viktor Savinykh
USSR

We left the spacecraft and entered outer space, and hurtling high above the Earth beside the vast and silent *Salyut* 7, we studied it attentively.

The solar panels were strangely oriented, their photoelectric cells peeling away in pieces, and looking for all the world like storm-torn sails. The once bright green shell had been burned and was now covered in grayish-rust spots. The portholes were all closed by shutters on the inside. No damage, however, could be seen on the hull of the craft. Well, what has happened to you, cosmic wanderer, we asked, but we got no answer.

Vladimir Dzhanibekov
USSR

Мы вышли из корабля в открытый космос и неслись над Землей вместе с замолкнувшей громадой «Салюта-7», внимательно ее изучая.

Развернутые странным образом, с отслоившимися кое-где фотоэлементами, солнечные панели напоминали потрепанные штормом паруса. Некогда ярко-зеленая обшивка выгорела и покрылась серовато-ржавыми пятнами. Все иллюминаторы изнутри закрыты шторками. Повреждений на корпусе не обнаружено. Что же с тобой случилось, космический странник? Ответа не последовало.

Владимир Джанибеков
СССР

Занятно, что в космическом корабле можно принять любое положение: вверх ногами, наискось под любым углом. Мне, например, нравилось прикреплять ноги к потолку, а Виталий Севастьянов «садился» на диван и фиксировал себя ремнями. Наши лица оказывались на одном уровне, но перевернутыми по отношению друг к другу на 180°. Нам так было удобно. Мы обедали, разговаривали, шутили. Со стороны, конечно, все выглядело более чем странно.

Андриян Николаев
СССР

It is amusing how any position, such as upside down or at an angle, is possible in a spacecraft. For instance, I like attaching my legs to the ceiling, while Vitali Sevastyanov "sat" strapped in the couch. Our heads turned out to be on the same level but twisted by 180 degrees. We found it convenient and we had dinner together and talked and joked with each other that way. Of course, from a distance it seems very strange.

Andriyan Nikolayev
USSR

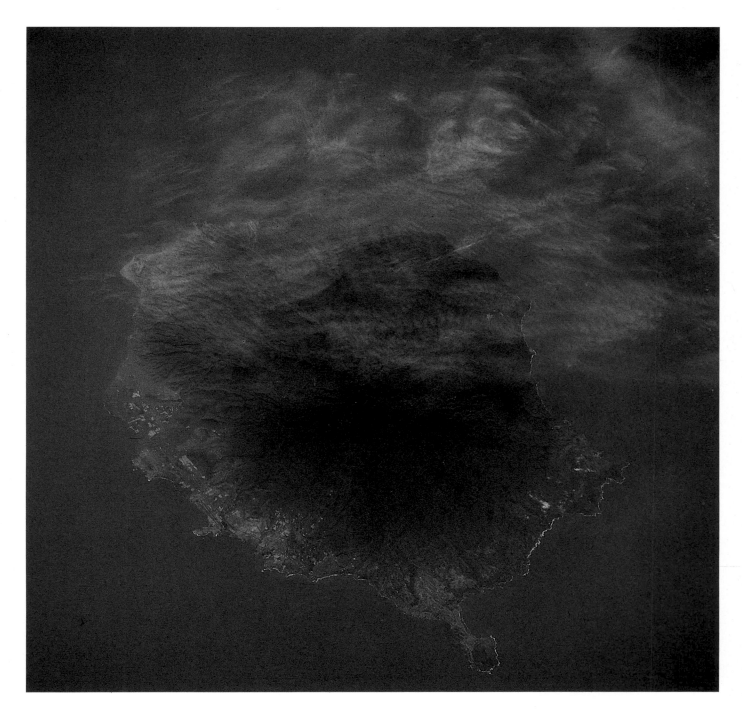

93   Gran Canaria Island, Canary Islands

Первые часы в космосе, конечно, не идиллия. Физическое ощущение такое, будто вся кровь устремилась к голове – голова тяжелая, при закрытых глазах кажется, что опрокидываешься назад.
Тело необыкновенно легкое, тренированные мышцы не имеют никакой нагрузки. Вестибулярный аппарат, который осуществляет ориентацию в пространстве – словно компас, стрелка которого неожиданно отклонилась от полюсов Земли. Сначала чувствуешь, что все время хочется держаться за что-то, держишься, отпускаешь со страхом и потом понимаешь, что некуда упасть и просто висишь в том же месте.

Георгий Береговой
СССР

The first hours in space are no idyll. You have the physical sensation of all your blood running to your head, which feels very heavy. When your eyes are closed it seems you are tumbling backwards. Either you are always floating up from somewhere or you are turning backward somersaults. There is an unusual lightness to your body, and your trained muscles seem to have no purpose. Your inner ear – the organ which gives you your sense of position – becomes like a compass whose pointer has suddenly lost the Earth's poles. To begin with, you feel you always want to hang onto something. You first hold on, letting go with trepidation, and then you find that there is nowhere to fall and that you simply hang in the same place.

Georgi Beregovoy
USSR

О'Генри, американский писатель, в одном из своих рассказов писал, что если вы хотите поощрять ремесло человеко-убийства, то нужно запереть двух мужчин на месяц в хижине 18 на 20 футов. Человеческая натура этого не выдержит.

Входя в Салют, который нам послужит и домом, и кабинетом шесть месяцев, мы сказали друг другу: «Мы братья. Я тебе, а ты мне.»

Валерий Рюмин
СССР

O. Henry, the American writer, wrote in one of his stories that if you want to encourage the craft of murder, all you have to do is lock up two men for two months in an eighteen-by-twenty-foot room.

Entering *Salyut,* which was to be both our home and our office for six months, we told each other: We are brothers. I am you and you are me.

Valeri Ryumin
USSR

87   Belcher Islands, Hudson Bay, Canada

Ngay từ thời thơ ấu tôi đã yêu bầu trời. Khi được bay, tôi đã muốn bay cao hơn nữa vào khoảng không gian bao la để trở thành "con người ở tầm cao". Sau 8 ngày bay trong vũ trụ tôi thấu hiểu được rằng, tầm cao cần thiết cho con người trước tiên là hiểu rõ hơn Trái Đất đã chịu nhiều đau khổ, quan sát tất cả những gì mà ở gần ta không thể nhận thấy. Không chỉ để chiêm ngưỡng vẻ đẹp kiều diễm của Trái Đất, mà còn để nhận rõ trách nhiệm, cần phải tiến hành công việc thế nào cho tốt hơn để không làm mảy may tổn hại tới thiên nhiên.

Phạm Tuân
Việt nam

I have been in love with the sky since birth. And when I could fly, I wanted to go higher, to enter space and become a "man of the heights." During the eight days I spent in space, I realized that mankind needs height primarily to better know our long-suffering Earth, to see what cannot be seen close up. Not just to love her beauty, but also to ensure that we do not bring even the slightest harm to the natural world.

Pham Tuan
Vietnam

86   Lake Van in eastern Turkey

As I looked down, I saw a large river meandering slowly along for miles, passing from one country to another without stopping. I also saw huge forests, extending across several borders. And I watched the extent of one ocean touch the shores of separate continents. Two words leaped to mind as I looked down on all this: commonality and interdependence. We are one world.

John-David Bartoe
USA

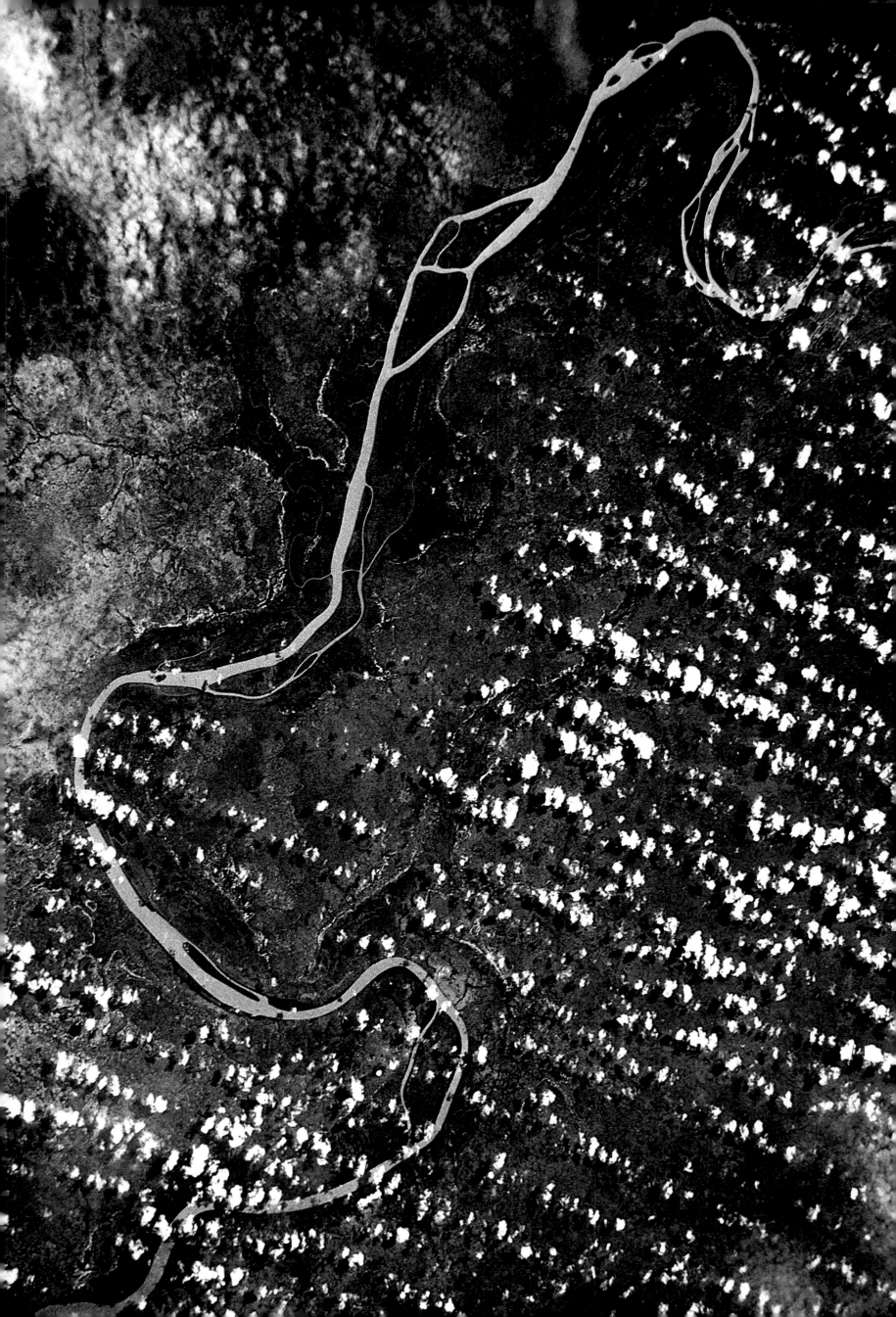

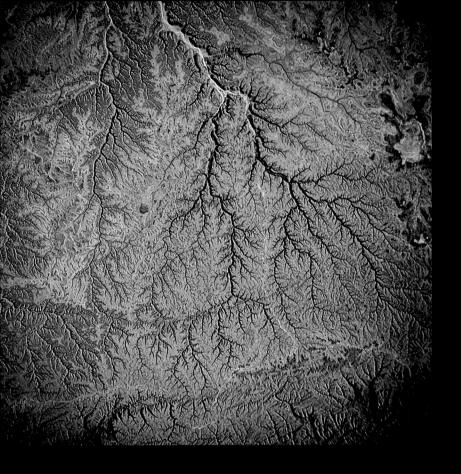

When you look out the other way toward the stars you realize it's an awful long way to the next wateri[n]g hole.

Loren Acton
USA

84   **Hadhramaut Plateau, South Yemen**

За 18 суток космического полета я убедился, что все обозримое пространство безжизненно. Черная пустота, белые немигающие звезды и планеты. Мысль об уникальности жизни и рода человеческого в безграничной Вселенной порой сильно угнетала меня, вызывала тоску и в то же время заставляла по-иному все осознать.

   Природа была беспредельно добра к нам, позволив нам, людям, появиться, выстоять, возмужать. Она щедро одарила нас всем тем, что скопила за миллиарды лет неодушевленного развития. Мы стали сильными и могущественными. А чем ответили на добро?

Юрий Глазков
СССР

After eighteen days of a space mission I was convinced that all visible space – the black emptiness, the white, unblinking stars and planets – was lifeless. The thought that life and humankind might be unique in the endless universe depressed me and brought melancholy upon me, and yet at the same time compelled me to evaluate everything differently.

   Nature has been limitlessly kind to us, having helped humankind appear, stand up, and grow stronger. She has generously given us everything she has amassed over the billions of years of inanimate development. We have grown strong and powerful, yet how have we answered this goodness?

Yuri Glazkov
USSR

Desde el espacio me ví simplemente como una persona más entre los millones y millones que han vivido, viven y vivirán sobre la Tierra. Inevitablemente, esto le hace a uno reflexionar sobre su propia existencia y la forma en que debemos vivir nuestras cortas vidas plenamente, disfrutando . . . compartiendo.

Rodolfo Neri-Vela
México

From space I see myself as one more person among the millions and millions who lived, live, and will live on Earth. Inevitably, this makes one think about our existence and the way in which we should live to enjoy, to share, our short lives as fully as possible.

Rodolfo Neri-Vela
Mexico

83   Parallel sand dunes, Algerian Desert

Tizenkét éves voltam akkor. A gyulaházi elemi iskolába jártam, Magyarország északkeleti részén. A gyulaházi parasztok, akik nagyon is földi gondolkodású emberek, úgy fogadták Gagarin űrutazásának hirét, mint valami természetfölötti dolgot. Sokan egyszerüen el sem akarták hinni, hogy az ember képes ilyen magasságokba szállni.

Farkas Bertalan
Magyar Népköztársaság

I was only twelve and at school in the village of Dyulahas in southeast Hungary when Gagarin flew. The peasants in the village were very down-to-earth people, and they greeted the news as something supernatural. Many were completely unwilling to believe that a man had been able to go so far from Earth.

Bertalan Farkas
Hungary

في اليوم الأول من مدارنا حول الأرض كان كل منا يشير إلى بلده عند مرورنا فوقها... ولكن مع مرور الأيام وعلى وجه التحديد في اليوم الثالث أو الرابع بدأ كل منا يشير إلى قارته... ومع قدوم اليوم الخامس لم نلاحظ حتى القارات وبدأنا ننظر إلى الأرض كوكب واحد.

سلطان بن سلمان آل سعود
المملكة العربية السعودية

The first day or so we all pointed to our countries. The third or fourth day we were pointing to our continents. By the fifth day we were aware of only one Earth.

Sultan Bin Salman al-Saud
The Kingdom of Saudi Arabia

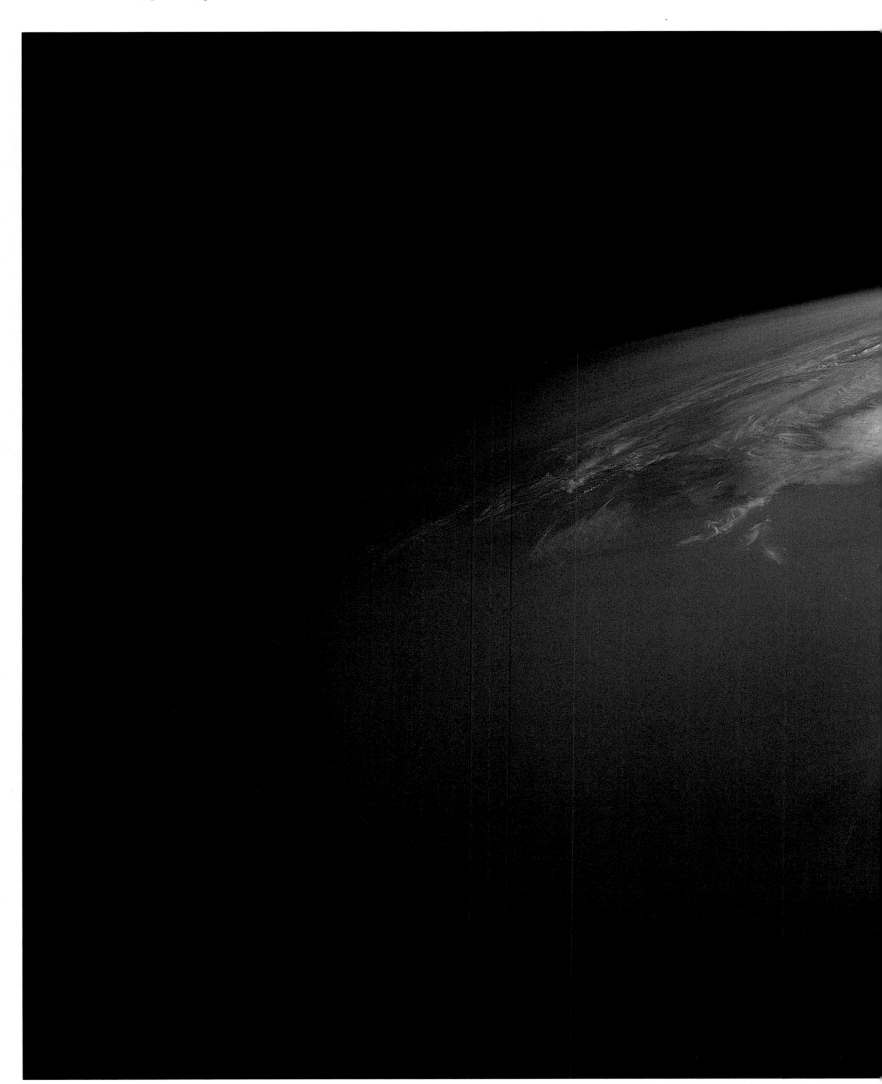

Vers six heures du soir nous avons survolé le littoral méditerranéen de la France presqu'au-dessus de Marseille. J'ai habité cette région pendant plus de vingt ans and je la connais bien. En même temps je pouvais voir toute la France, la Corse, la Sardaigne, l'Italie, une partie de l'Espagne, discerner le sud de l'Angleterre, une partie de l'Allemagne. En somme, je voyais une partie considérable de la Terre, tout en distinguant sans difficulté les petits détails du terrain où j'avais marché à pied quelques semaines auparavant. Alors je souriais de me rendre compte de l'immensité dérisoire et relative de notre planète. Quelques secondes plus tard, nous survolions l'URSS!

Jean-Loup Chrétien
France

Around six pm we were flying north of the Mediterranean, almost right above Marseilles. I know this region well, since I lived there more than twenty years. At the same time I can see all of France, Corsica, Sardinia, Italy, part of Spain; perceive the south of England, part of Germany, in fact, a good part of the world; all the while distinguishing without difficulty the little details of the place where I was wandering on foot some weeks earlier. I smile then, realizing how laughable and relative the immensity of our planet is. Some seconds later, we were flying above the USSR!

Jean-Loup Chrétien
France

धीरे धीरे मेरा मानसिक क्षितिज विस्तृत होता गया... यह भावना और भी प्रगाढ हो गई जब पृथ्वी को मैंने एक सीमारहित, आकर्षणहीन काले शून्य की पृष्ठभूमि में देखा । मुझे इस बात की प्रसन्नता थी कि मेरे देश की सदियों पुरानी समृद्ध परम्पराओं एवं संस्कृति ने मुझे मनुष्य द्वारा बनाई हुई सीमाओं और पक्षपातों से परे देखने के योग्य बनाया है । मुझे यह भी अनुभव हुआ कि एकत्व की इस भावना को महसूस करने के लिए, किसी को अंतरिक्ष में जाने की आवश्यकता नहीं ।

राकेश शर्मा
भारत

My mental boundaries expanded when I viewed the Earth against a black and uninviting vacuum, yet my country's rich traditions had conditioned me to look beyond man-made boundaries and prejudices. One does not have to undertake a space flight to come by this feeling.

Rakesh Sharma
India

The clouds were always different, the light was different. Snow would fall, rain would fall – you could never depend on freezing any image in your mind.

William Pogue
USA

One expects blues and greens, whites and browns, but the pinks, purples, yellows! I can only think about the colors that one experiences seeing our planet for the first time from space.

Byron Lichtenberg
USA

78   Linear clouds over Argentina

Проходим над Гималаями. Видим хребты с высочайшими вершинами мира. На краю долины Катманду, вытянувшейся с севера на юг, нашел Эверест. Как много людей мечтает взобраться на него, чтобы оттуда взглянуть вниз, а нам его сверху даже трудно распознать.

Валентин Лебедев
СССР

We are passing over the Himalayas. We can see the mountain ranges with the highest peaks in the world. At the end of the Katmandu valley, which runs from north to south, I found Everest. How many people dream of conquering Everest, so that they can look down from it, and yet for us from above it was difficult even to locate it.

Valentin Lebedev
USSR

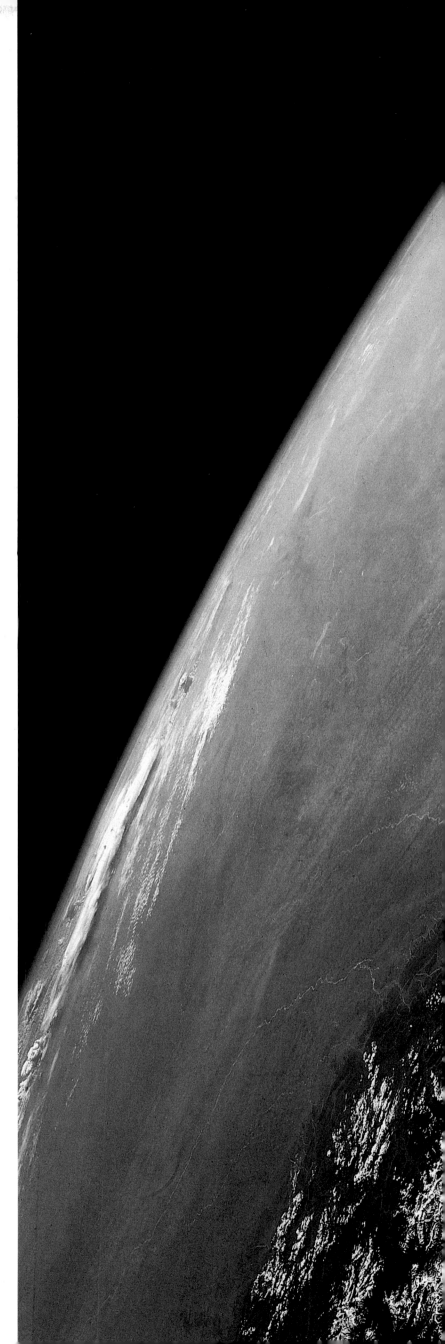

После того, как оранжевое облако,
образовавшееся в результате
пыльной бури над Сахарой и
подхваченное воздушными
течениями, достигло Филиппинских
островов и осело там с дождем, мне
стало понятно, что все мы плывем в
одной лодке.

Владимир Коваленок
СССР

After an orange cloud – formed as a
result of a dust storm over the Sahara
and caught up by air currents –
reached the Philippines and settled
there with rain, I understood that we
are all sailing in the same boat.

Vladimir Kovalyonok
USSR

من الفضاء رأيت الأرض جميلة رائعة وتلاشت
الحدود بين البلدان.

محمد أحمد فارس
سوريا

From space I saw Earth – indescribably
beautiful with the scars of national
boundaries gone.

Muhammad Ahmad Faris
Syria

76   Dasht-i-Kavir, Iran

Africa looked ill with its sandstorms and the dried-out areas.

Robert Overmyer
USA

Madagascar is still green with tropical forest but probably not for long. The ocean around that island is colored a thick bloody red by the silt that is being eroded from recently deforested areas.

And how much desert there is on the Earth. Over Africa I never saw a great expanse of green tropical rain forest. North Africa is the Sahara Desert, Southwest Africa is the Namib Desert, and South and East Africa is semi-arid grassland.

Karl Henize
USA

The signs of life are subtle but unmistakable: sprawling urban concentrations, circular irrigation patterns, the wakes of ships, bright city lights at night, and burning oil fields.

Marc Garneau
Canada

The sunlight on seemingly "dry" land, in semi-arid South and East Africa, for example, has a special beauty. Myriad small ponds and streams would reflect the full glare of the sun for one or two seconds, then fade away as a new set of water surfaces came into the reflecting position. The effect was as if the land were covered with sparkling jewels.

Karl Henize
USA

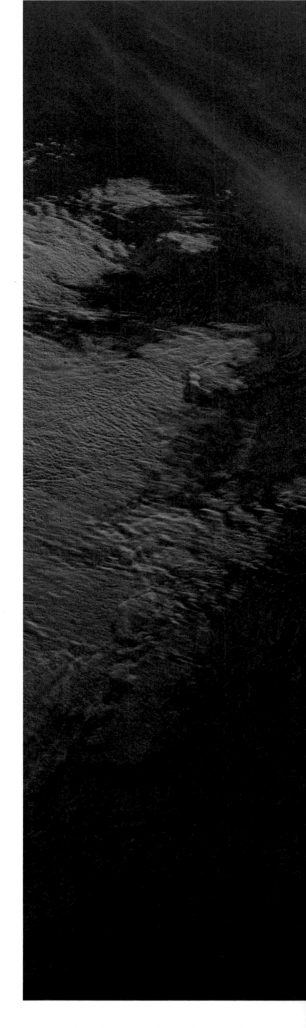

73   Glacial lakes in the Canadian Shield of Quebec

72   Irrigation patterns in Brazil

Ich mache mir Sorgen, wenn der russische Kosmonaut mir erzählt, daß die Atmosphäre über dem Baikal-See genauso verschmutzt ist wie über Europa, und wenn der amerikanische Austronaut erwähnt, daß er vor 15 Jahren viel klarere Fotografien von Industriezentren aufnehmen konnte als heute.

Ernst Messerschmid
Bundesrepublik Deutschland

When the Russian cosmonaut tells me that the atmosphere over Lake Baikal is as polluted as it is over Europe, and when the American astronaut tells me that fifteen years ago he could take much clearer pictures of the industrial centers than today, then I am getting concerned.

Ernst Messerschmid
Federal Republic of Germany

70 Indonesia

Неважно, в каком озере или море
ты обнаружил очаги загрязнения
или в лесах какой страны увидел
возникшие очаги пожаров, над каким
континентом зарождается ураган. Ты
охраняешь всю свою Землю.

Юрий Артюхин
СССР

It isn't important in which sea or lake
you observe a slick of pollution, or in
the forests of which country a fire
breaks out, or on which continent a
hurricane arises. You are standing
guard over the whole of our Earth.

Yuri Artyukhin
USSR

Un cosmonaut, dacă ar mînca undeva
lîngă planeta Marte, pîine coaptă din
grîu crescut într-un laborator cosmic,
credeți-mă, că și acolo, el s-ar gîndi la
spicele de grîu și florile culese pe
Pămînt!

Dumitru Prunariu
România

A cosmonaut, should he eat bread
somewhere near Mars baked from
wheat raised in a space laboratory,
will, believe me, think of the grain and
flowers gathered on Earth!

Dumitru Prunariu
Romania

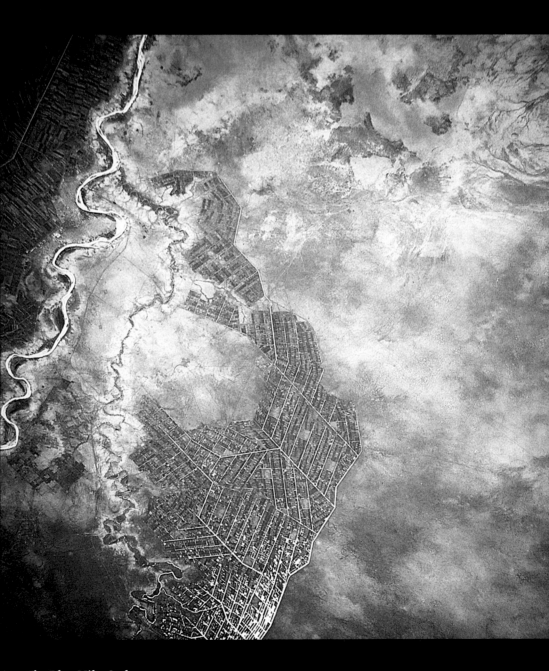

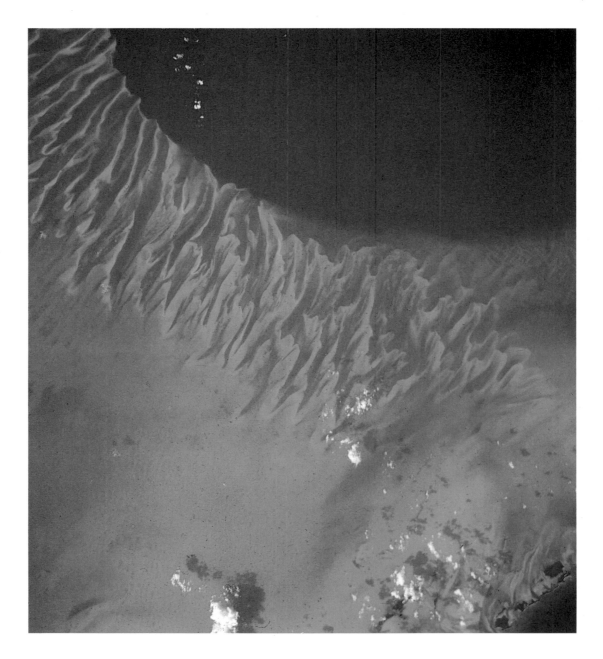

Пролетая над Мозамбикским
проливом, разделяющим
африканский континент и остров
Мадагаскар, хорошо видели
поперечные песчаные отмели на его
дне. Совсем как речушка, которую
переходил вброд в детстве.

Лев Дёмин
СССР

As we were flying over the
Mozambique Channel, which
separates the island of Madagascar
from the continent of Africa, we
could clearly see the transverse sand
bars at its bottom. It was just like a
brook one waded in in childhood.

Lev Demin
USSR

In the middle of the night we were lying on the ceiling looking out the window at the most fantastic thing that any human being has ever seen. Then we both said it almost at the same time – that the only sadness to the whole experience was the fact that there was the very real possibility that our wives and children would never share that with us.

Robert Cenker
USA

Now as we were crossing the eastern coast of South America I looked out to the sea and got an absolutely perfect view of the confluence of the Falkland and South Equatorial currents. You can see the Falkland Current coming up from the south, and you can see, not quite so clearly, the Equatorial Current coming down from the north. And out from the city of Buenos Aires you can see where these two currents meet and head straight to the southeast.

Gerald Carr
USA

66   A plankton bloom off New Zealand

We were able to see the plankton blooms resulting from the upwelling off the coast of Chile. The bloom itself extended along the coastline and had some long tenuous arms reaching out to sea. The arms or lines of plankton which were pushed around in a random direction, fairly well defined but fairly weak in color, contrasted with the dark blue ocean. The fishing ought to be good down there.

Edward Gibson
USA

Поверхность океана сначала кажется совершенно однообразной; однако через полмесяца мы начали различать по характерным оттенкам различные моря и даже части океанов. Мы с удивлением обнаружили, что во время полета заново приходится учиться не только смотреть, но и видеть. Сначала тонкие цветовые оттенки ускользают от глаза, но постепенно зрение как будто обостряется, цветовое восприятие становится богаче и, наконец, вся планета предстает перед глазами во всей своей неописуемой красоте.

Владимир Ляхов
СССР

Although the ocean's surface seems at first to be completely homogeneous, after half a month we began to differentiate various seas and even different parts of oceans by their characteristic shades. We were astonished to discover that, during a flight, you have to learn anew not only to look, but also to see. At first the finest nuances of color elude you, but gradually your vision sharpens and your color perception becomes richer, and the planet spreads itself before you with all its indescribable beauty.

Vladimir Lyakhov
USSR

The Pacific. You don't comprehend it
by looking at a globe, but when you're
traveling at four miles a second and it
still takes you twenty-five minutes to
cross it, you know it's big.

Paul Weitz
USA

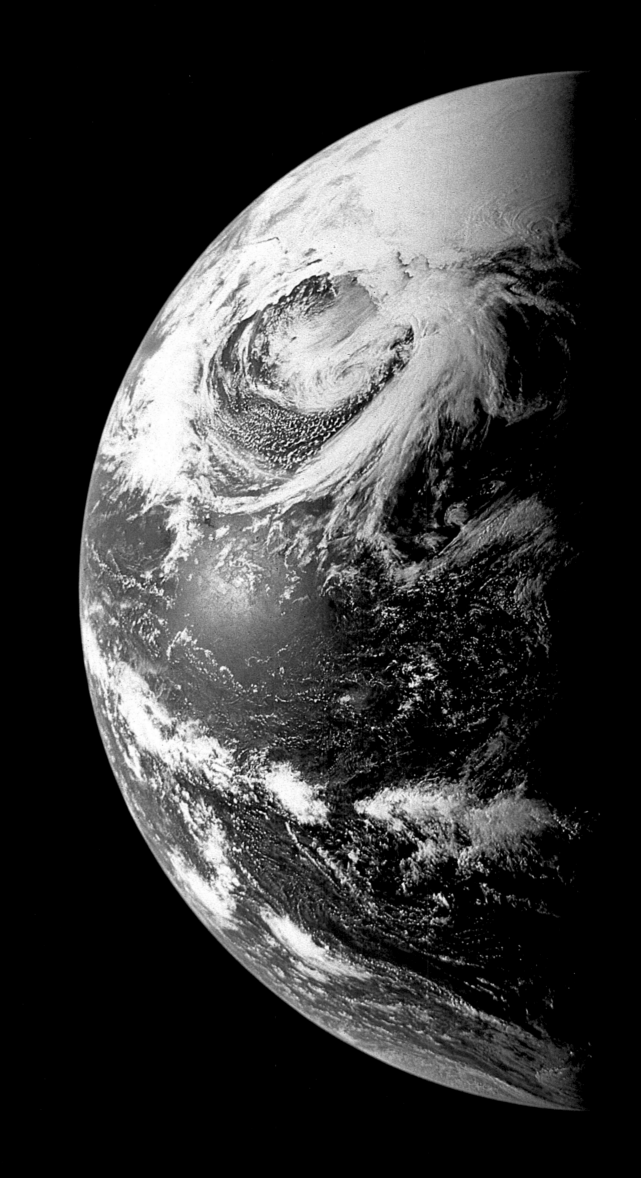

63   The western coast of Australia

Cette beauté est faite de nuances
subtiles, d'un équilibre miraculeux
de teintes resplendissantes et douces.
Seul un enfant dans son innocence
pourrait appréhender la pureté et la
splendeur de cette vision.

Patrick Baudry
France

This beauty consists of subtle nuances,
as in the miraculous balance of soft
and brilliant hues. Only a child in its
innocence could apprehend the purity
and splendor of this vision.

Patrick Baudry
France

62   The Gulf of Carpentaria, Australia

Однако ловишь себя на мысли, что
смотришь на всю эту красоту под
профессиональным углом зрения:
отмечаешь степень прозрачности
атмосферы и метеоструктуру, следы
течений в океане, скопления
планктона, техногенные загрязнения,
интересные геологические структуры,
места лесных пожаров, участки
засоленности почв,
идентифицируешь, фотографируешь,
записываешь – в общем, работаешь.
А время то стремительно летит,
обгоняя корабль, то почти стоит –
совсем по Эйнштейну.

Юрий Романенко
СССР

You realize that you are looking at all
this beauty with a professional eye:
You note the transparency of the
atmosphere and its meteorological
structure, the traces of the currents in
the oceans, accumulations of
plankton, concentrations of industrial
pollution, interesting geological
structures, areas where forest fires had
raged, stretches of saline soil; you
identify, photograph, jot things down;
basically you work. Sometimes time
flies, and sometimes it almost stands
still; like Einstein said, everything is
relative.

Yuri Romanenko
USSR

Пролетая над планетой, мы
наблюдали некоторые непонятные,
таинственные явления, например,
появляющееся перед восходом солнца
странное мерцающее свечение
вблизи экватора, похожее на северное
сияние, и коричневые, непрерывно
меняющиеся тени на дневной стороне
Земли.

Валерий Рюмин
СССР

Flying above the planet, we have
observed certain incomprehensible
phenomena – for example, the strange
flickering luminescence before
sunrise near the equator, which
resembles the northern lights, and
also the brown shadows on the day
side of the Earth that are constantly
changing.

Valeri Ryumin
USSR

They say if you have experiments to run, stay away from the window. For me, preoccupied with the Drop Dynamics Module, it wasn't until the last day of our flight that I even had a chance to look out. But when I did, I was truly overwhelmed.

A Chinese tale tells of some men sent to harm a young girl who, upon seeing her beauty, become her protectors rather than her violators. That's how I felt seeing the Earth for the first time. "I could not help but love and cherish her."

Taylor Wang
China/USA

美國／中國　王贛駿

地球美景
我見猶憐

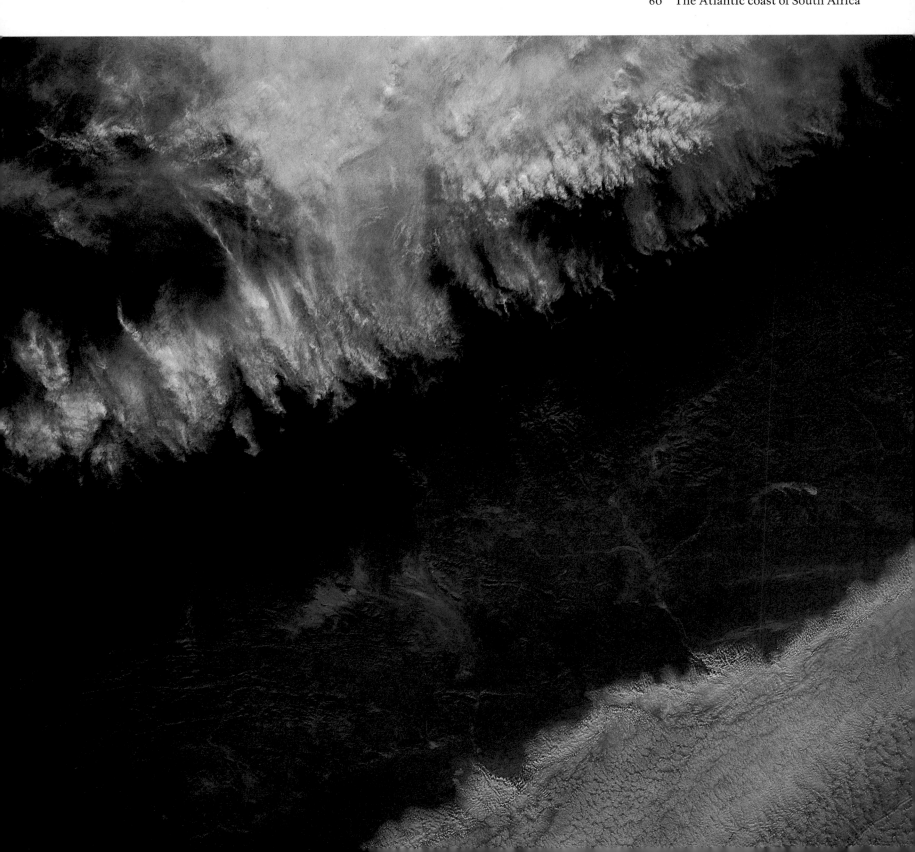

You see layers as you look down. You
see clouds towering up. You see their
shadows on the sunlit plains, and you
see a ship's wake in the Indian Ocean
and brush fires in Africa and a
lightning storm walking its way
across Australia. You see the reds and
the pinks of the Australian desert, and
it's just like a stereoscopic view of all
nature, except you're a hundred ninety
miles up.

Joseph Allen
USA

Zum ersten Mal in meinem Leben
sah ich den Horizont als eine
gobogene Linie. Sie war durch eine
dunkelblaue dünne Naht betont-
unsere Atmosphäre. Offensichtlich
handelte es sich hierbei nicht um das
Luftmeer, wie man mir oft in meinem
Leben erzählte. Die zerbrechliche
Erscheinung versetzte mich in
Schrecken.

Ulf Merbold
Bundesrepublik Deutschland

For the first time in my life I saw
the horizon as a curved line. It was
accentuated by a thin seam of dark
blue light – our atmosphere. Obviously
this was not the ocean of air I had been
told it was so many times in my life. I
was terrified by its fragile appearance.

Ulf Merbold
Federal Republic of Germany

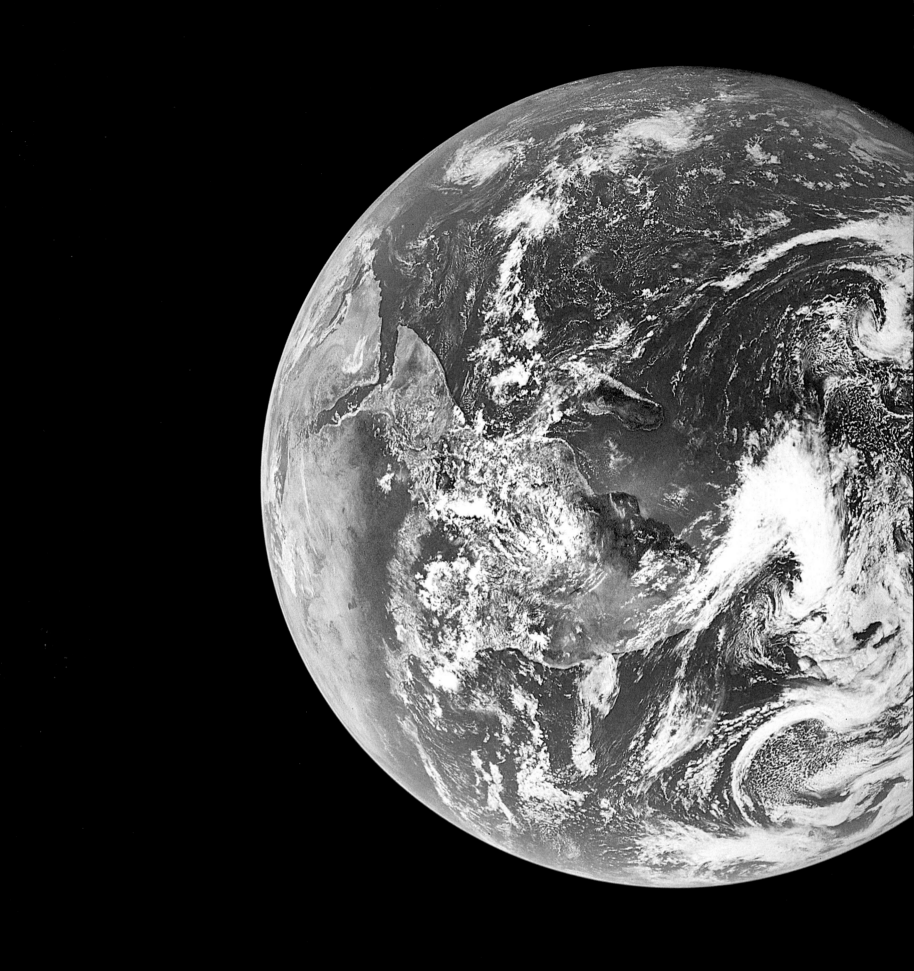

Observations

Через несколько суток созерцания Земли возникла ребяческая мысль, что нас обманывают: «Если мы первенцы в космосе, то кто же смог так правильно сделать глобус?», сменившаяся гордостью за способность человека видеть разумом.

Игорь Волк
СССР

Several days after looking at the Earth a childish thought occurred to me – that we the cosmonauts are being deceived. If we are the first ones in space, then who was it who made the globe correctly? Then this thought was replaced by pride in the human capacity to see with our mind.

Igor Volk
USSR

Мне кажется, что даже мудрейшие философы Возрождения, самые дерзкие умы прошлого не могли реально оценить масштабы нашей планеты. Раньше она казалась огромной, почти бесконечной. Только в шестидесятые годы нашего столетия человек, поднявшись над Землей в космос, с удивлением и некоторым разочарованием обнаружил, насколько в действительности она мала. Некоторым она показалась островом в безграничном океане мироздания. Другие сравнивали ее с космическим кораблем, экипаж которого превышает шесть миллиардов человек.

Павел Попович
СССР

It seems to me that even the wisest philosophers of the Renaissance or the most daring minds from the past could not estimate the real size of our planet. Earlier, it seemed immeasurably great, almost infinite. Only after the middle of this century did man, having gone up above the Earth into space, see with surprise and disbelief just how small the Earth really is. Some saw it as an island in the limitless ocean of creation. Some compared it to a spaceship with a crew numbering more than six billion.

Pavel Popovich
USSR

We spent most of the way home
discussing what color the moon was.

Eugene Cernan
USA

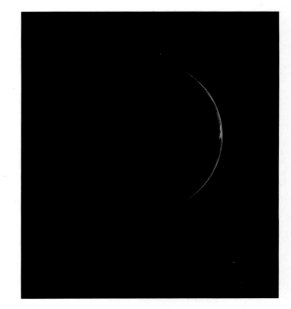

On the way back we had an EVA. I had
a chance to look around while I was
outside and Earth was off to the right,
180,000 miles away, a little thin sliver
of blue and white like a new moon
surrounded by this blackness of space.
Back over my left shoulder was almost
a full moon. I didn't feel like I was a
participant. It was like sitting in the
last row of the balcony, looking down
at all of that play going on down there.
That was really the only time that I
had that "insignificant" feeling of the
immensity of this, God's creation.

Charles Duke, Jr.
USA

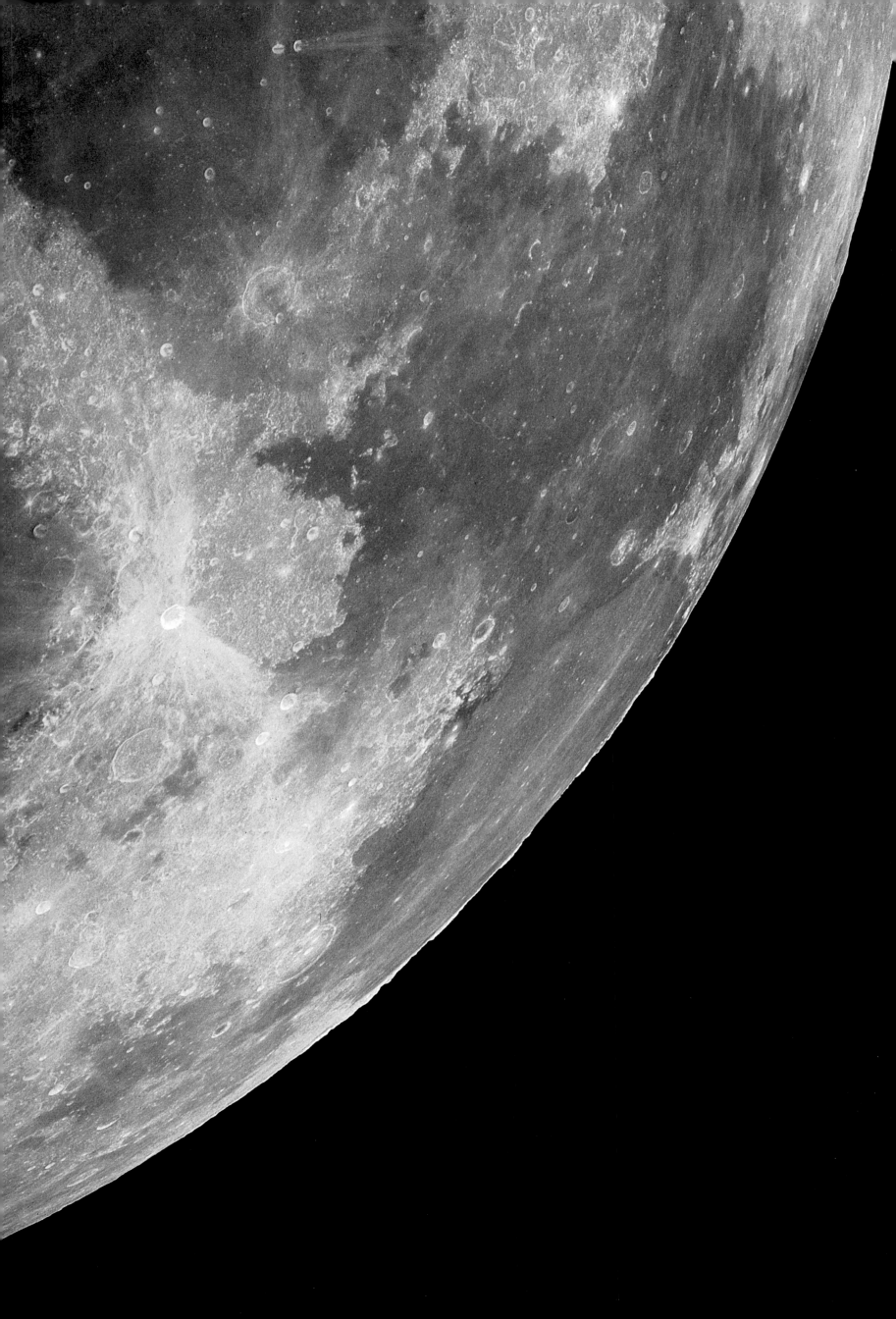

I enjoyed the fact that I was on one side of this little satellite of our planet and Neil and Buzz were somewhere over on the other side of it, and then there were three billion people a quarter of a million miles away. And that was all over there, and if I looked in the other direction there was God knows what: only me and the rest of the universe. I liked that feeling, being a part of the rest of the universe instead of part of the Solar System. I didn't mind being in that corner of the universe alone by myself. I enjoyed that. I wish someone would have communicated with me, but no one did.

Michael Collins
USA

You look out the window and you're looking back across blackness of space a quarter of a million miles away, looking back at the most beautiful star in the heavens. You're not close enough to any other planets to see anything but a bright star, but you can look back on the Earth and see from pole to pole and across oceans and continents, and you can watch it turn and see there are no strings holding it up, and it's moving in a blackness that is almost beyond conception.

The Earth is surrounded by blackness though you're looking through sunlight. There is only light if the sunlight has something to shine on. When the sun shines through space it's black. All because the light doesn't strike anything. The light doesn't strike anything, so all you see is black.

What are you looking at? What are you looking through? You can call it the universe, but it's the infinity of space and the infinity of time.

Eugene Cernan
USA

53   The Earth eclipsing the sun

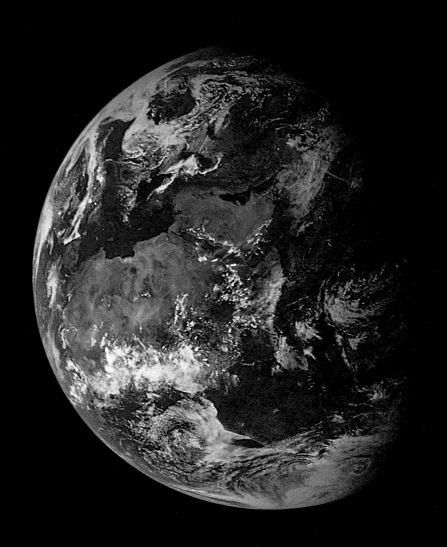

One thing that I remember most visually was being on the moon and looking back at the Earth and thinking how far, far away it was. I was impressed by the distance; it seemed very unreal for me to be there. Frequently on the lunar surface I said to myself, This is the moon, that is the Earth. I'm really here. I'm really here.

Alan Bean
USA

My view of our planet was a glimpse of divinity.

Edgar Mitchell
USA

Now I know why I'm here
Not for a closer look at the moon,
But to look back
At our home
The Earth.

Alfred Worden
USA

I have always thought it was curious that on the moon all the stars circulate around you, but not as fast as they do here; they do it once every twenty-eight days instead of once every twenty-four hours, and the sun moves around you the same way. Yet the Earth, which is the biggest object there, stays right in the same spot. I thought, If ancient man had been born on the moon instead of the Earth, he would have had much more difficulty determining what was going on, because these things would be in slow motion except for the one which was still. I felt pretty sure that in ancient cultures they would have worshipped the Earth and thought it was an eye, because it would change from blue to white and you would see something moving up there that did look like a colored eye.

Alan Bean
USA

51 The Earth viewed from *Apollo 15*

As I orbited the moon, and the moon
was in turn making its twenty-eight-
day orbit around the Earth, I could
watch the Earth change from three-
quarters to one-half and on down to a
crescent.

Ronald Evans
USA

I felt like I was an alien as I traveled through space. When I got on the moon, I felt at home. We had mountains on three sides and had the deep canyon to the west, a beautiful spot to camp. I felt in a way as Adam and Eve must have felt, as they were standing on the Earth and they realized that they were all alone.

I talk about the moon as being a very holy place.

James Irwin
USA

46   Boulders on the moon

47   The valley of Taurus-Littrow on the moon

42-45   Earthrise as seen from the moon

Suddenly from behind the rim of the moon, in long, slow-motion moments of immense majesty, there emerges a sparkling blue and white jewel, a light, delicate sky-blue sphere laced with slowly swirling veils of white, rising gradually like a small pearl in a thick sea of black mystery.

It takes more than a moment to fully realize this is Earth . . . home.

Edgar Mitchell
USA

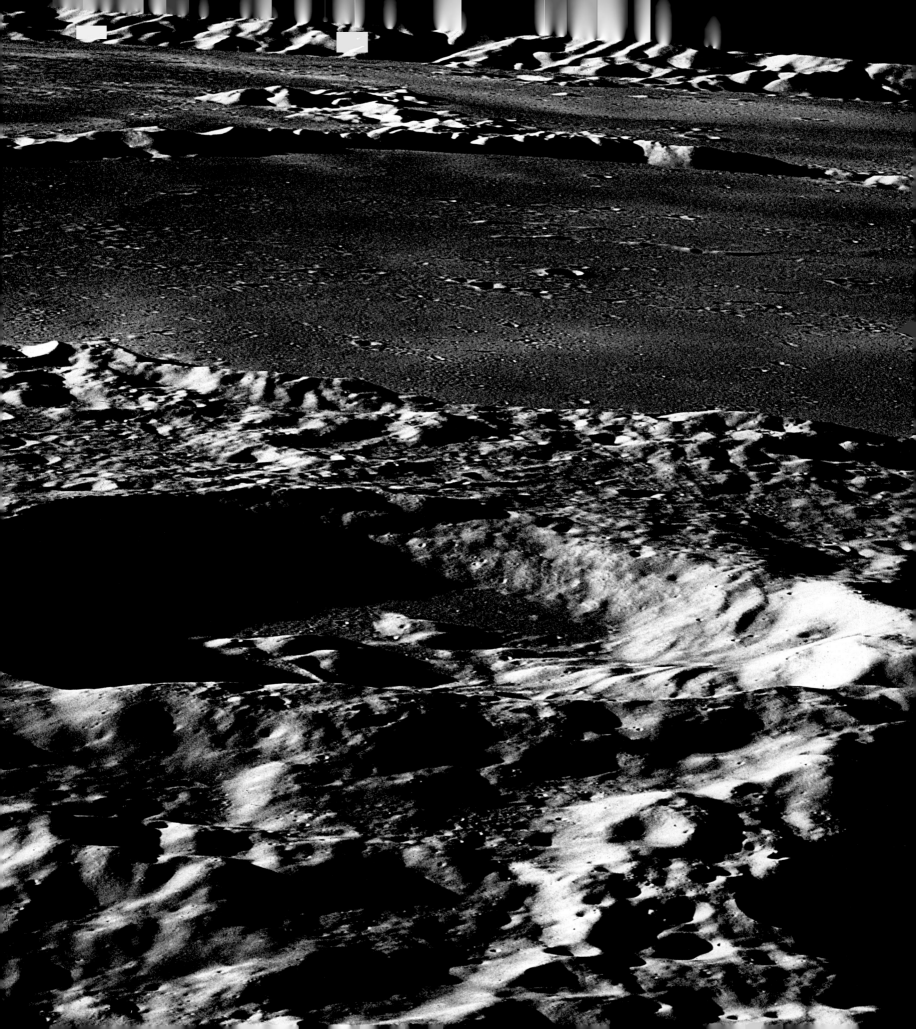

I was lying there, looking out the window as we moved across the terminator. I was listening to the *Symphonie Fantastique,* and it was dark in the spacecraft. I was looking down at dark ground, and there was Earthshine. It was like looking at a snow-covered Earth scene under a full moon.

Ken Mattingly
USA

We were in its shadow, and we were seeing its surface with sunlight that had traveled from the sun and then bounced off the surface of the Earth and then back up to the moon.

Michael Collins
USA

On the backside of the moon, on the night side, you can't see the surface. The moon is defined simply by the absence of stars. The laws of physics tell you that your fine spacecraft is in an orbit sixty miles above it and there's no way you can hit anything. But the thought does occur, Gosh, I'm skimming along just barely over the surface of a strange planet.

Michael Collins
USA

It was a totally different moon than I had ever seen before. The moon that I knew from old was a yellow flat disk, and this was a huge three-dimensional sphere, almost a ghostly blue-tinged sort of pale white. It didn't seem like a very friendly or welcoming place. It made one wonder whether we should be invading its domain or not.

Michael Collins
USA

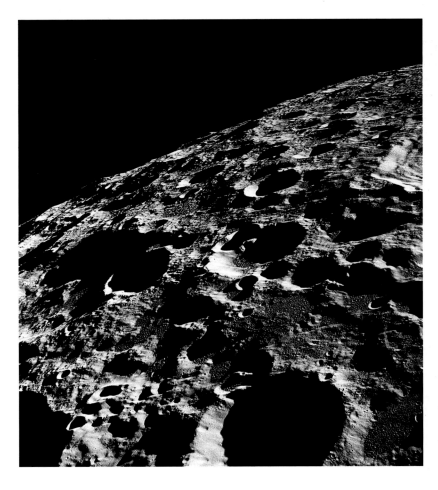

40   Craters on the far side of the moon

You see the sun come around every so often as you rotate the spacecraft. Then you see the moon coming around, so through the windows of the spacecraft you get this constant parade of darkness and stars on one side, and then the Earth swings through, and then the sun swings through, and then the moon swings through, and back to the star-filled skies again. It's eerie. You suddenly start to recognize you're in deep space, that planets are just that, they're planets, and you're not really connected to anything anymore. You are floating through this deep black void.

Edgar Mitchell
USA

The Earth reminded us of a Christmas tree ornament hanging in the blackness of space. As we got farther and farther away it diminished in size. Finally it shrank to the size of a marble, the most beautiful marble you can imagine. That beautiful, warm, living object looked so fragile, so delicate, that if you touched it with a finger it would crumble and fall apart. Seeing this has to change a man, has to make a man appreciate the creation of God and the love of God.

James Irwin
USA

37 The coast of Africa

We could not immediately detect the fact that the Earth was shrinking as we sped away from it. The sensation was rather like watching the hand on a clock move. You know it is moving, but watching it you cannot see it move. Only after looking elsewhere for a time, then returning to the minute hand, can one realize it actually did move. The Earth would eventually be so small I could blot it out of the universe simply by holding up my thumb.

Buzz Aldrin
USA

Under a full moon the clouds are luminous in pearly reflection. The high-altitude atmospheric airglow appears on the horizon as a stunning bronze-colored band above the now dark air. And lightning presents a show of beauty and power as it illuminates billowing clouds and weather fronts sometimes for distances of tens of miles. Beneath this natural wonder cities glow yellow or white, but are diminutive in scale.

Charles Walker
USA

There seems to be some sort of a collective organization to the lightning. When one goes off, two or three may go off simultaneously, or one of those may turn out to trigger a whole lot of other ones all over a very, very wide area – 500,000 square miles, perhaps.

Edward Gibson
USA

32-34   Moonrise as seen from *Skylab*

The full moon's ri
sunrise. As it com
horizon, it appear
similar to moonri
However, the risi
looks like a bubbl
water. As the moo
above the horizon
from a flattened b

William Pogue
USA

Na tmavém pozadí Země se rozžíhaly doutnavky meteoritů, občas se skoro strašidelně zablýsklo. Oddávám se čarovné síle světelných efektů a cítím se jako malý chlapec, který s pootevřenými ústy obdivuje slavnostní ohňostroj. Zničehonic se přede mnou objevuje něco kouzelného. Od Země, téměř k naší stanici (orbitální) se rozlévá nazelenalá záře, připomínající ohromné, fosforeskující píšťaly varhan, jejichž konce přechází do sytě karmínové barvy. Záře se prolíná trhanci jakoby válicí se páry.

Můžeš si gratulovat, Vladimíre – říkám si pro sebe. Měl si štěstí pokochat se nádherou polární záře.

Vladimír Remek
ČSSR

Firefly meteorites blazed against a dark background, and sometimes the lightning was frighteningly brilliant. Like a boy, I gazed open-mouthed at the fireworks, and suddenly, before my eyes, something magical occurred. A greenish radiance poured from Earth directly up to the station, a radiance resembling gigantic phosphorescent organ pipes, whose ends were glowing crimson, and overlapped by waves of swirling green mist.

Consider yourself very lucky, Vladimir, I said to myself, to have watched the northern lights.

Vladimir Remek
Czechoslovakia

Я вздрогнул, когда вместо привычного звездного неба в иллюминаторе на ночном горизонте планеты увидел багровое пламя. Огромные столбы света рвались в поднебесье, таяли в нем, переливали всё всеми цветами радуги. Область красного свечения плавно переходила в черный цвет космоса. Интенсивные и динамичные изменения цвета и формы столбов и гирлянд напоминало цветомузыку. Наконец, мы отчетливо увидели, что входим прямо в полярное сияние.

Александр Иванченков
СССР

I shuddered when I saw a crimson flame through the porthole instead of the usual starry sky at the night horizon of the planet. Vast pillars of light were bursting into the sky, melting into it, and flooding over with all the colors of the rainbow. An area of red luminescence merged smoothly into the black of the cosmos. The intense and dynamic changes in the colors and forms of the pillars and garlands made me think of visual music. Finally, we saw that we were entering directly into the aurora borealis.

Aleksandr Ivanchenkov
USSR

I somehow pictured always knowing where the Earth would be, even in darkness. Either there were going to be lights I could see on the ground, or there would be light leaking over the horizon from a soon-to-be-rising sun. Instead, I found the blackest black I ever saw.

The way you find the Earth in the dark is to track the stars until the stars stop. When the stars stop, that's the Earth blocking their light.

Joseph Allen
USA

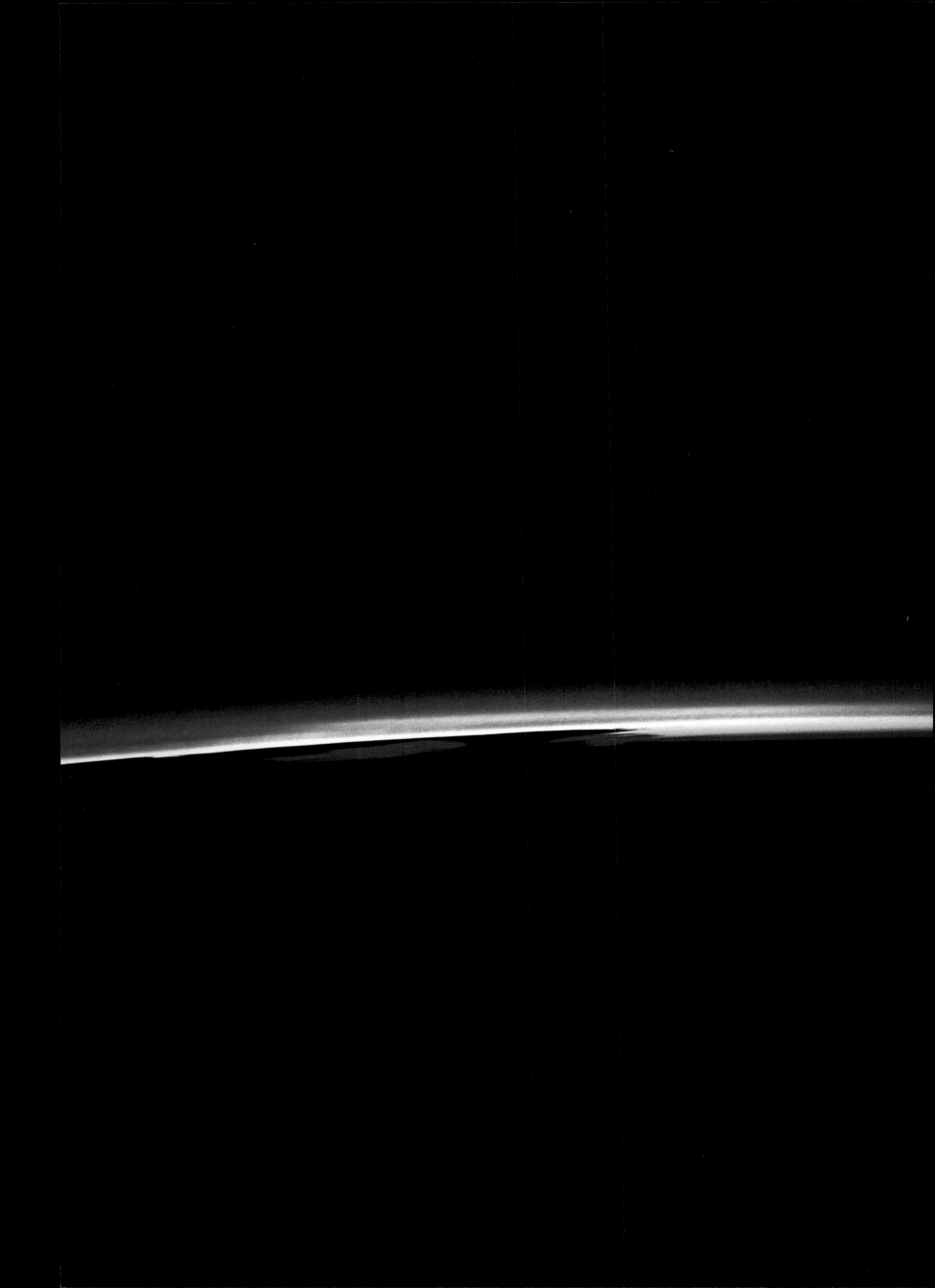

Второй выход в космос не был
таким напряженным; можно было
рассматривать Землю, особенно когда
мы находились в тени и у нас не
было дел.

 Мы глядели на звезды, видели,
как восходит и заходит Луна, как
проплывают под нами на Земле
континенты. Похоже было на то, как
если бы мы лежали где-нибудь на
пляже или лужайке и наблюдали за
проплывающими в небе облаками.

Владимир Ляхов
СССР

The second walk was not so stressful,
and we had time to look at Earth when
we were in a shadow and there was no
pressing work to do. We looked at the
stars, looked at the moon rising and
setting, and at the continents floating
beneath us. We compared the
experience with lying on the beach or
on the grass and looking up from Earth
at the passing clouds in the sky.

Vladimir Lyakhov
USSR

I had given up trying to use a small
flashlight to continue our work in the
dark. I raised the visor on my helmet
cover and looked out to try to identify
constellations. As I looked out into
space, I was overwhelmed by the
darkness. I felt the flesh crawl on my
back and the hair rise on my neck. I
was reminded of a passage in the Bible
that speaks of the "horror of great
darkness." Ed and I pondered the view
in silence for a few moments, and then
we both made comments totally
inadequate to describe the profound
effect the scene had made on both of
us. "Boy! That's what I call dark."

William Pogue
USA

It was a texture. I felt like I could reach
out and touch it. It was so intense. The
blackness was so intense.

Charles Duke, Jr.
USA

Literally out in space on *Gemini 9*,
walking twice around the world, there
was a point when I stopped and I
looked down at the world and I looked
around me and I realized that I was
really out in the pure vacuum of space
– really there, right that very instant.

Eugene Cernan
USA

В 6 ч. 20 мин. мы вошли в тень. Люк
на станции блестел как открытая
дверь деревенского дома.

Валентин Лебедев
СССР

At 0620 hours we went into the shade.
The station hatch shone like an open
door in a house in the countryside.

Valentin Lebedev
USSR

Я втиснулся обратно в переходный отсек. Юра Романенко все это время страховал меня изнутри: держал за ноги, чтобы я не улетел в космос. И тут он говорит: «Что ж я так и просидел в отсеке, увидел Землю и звезды только через стекло? Как они там выглядят.» Я посторонился, пропустил его к выходному люку. Он тут же рванулся мимо эдакой торпедой. Уже наружу выходит, как вдруг смотрю – вслед за ним плывет, извиваясь змейкой, страховочный фал, ни к чему не пристегнутый. Я успел фал подхватить и спрашиваю: «Ты куда лететь-то собрался?»

Георгий Гречко
СССР

I squeezed back into the intermediate module. Yuri Romanenko had been protecting me all the time from the inside, holding me by the legs so that I wouldn't float off into space. Then he said, "How come I'm stuck here looking at the Earth and the stars through glass? Let me see what they look like out there!" I moved aside and let him through to the exit lock. He jerked past me like a torpedo. He was already outside when suddenly I saw his safety line wriggling like a snake as it trailed after him attached to nothing. I grabbed hold of it and asked him, "Where do you think you're flying off to?"

Georgi Grechko
USSR

Ослепительно яркий свет в космосе оказался для меня неожиданным. Алексей Леонов говорил мне, что солнце в космосе светит очень ярко, но когда я вышел в открытый космос и посмотрел на наш космический корабль, я забыл, что смотрю через светопоглощающие очки. Так ослепительно сиял в лучах солнца наш корабль. Когда я сдвинул защитные очки (вообще говоря, это не допускалось, но соблазн был слишком велик), солнце было справа от меня и светило так ярко, что в ту сторону было невозможно смотреть.

Евгений Хрунов
СССР

The tremendous brightness of light in space came as a surprise to me. Aleksei Leonov had told me that the sun was very bright there, but when I floated out and looked at the ship it was so dazzling I forgot that I was looking through a dense filter. When I lifted the filter (strictly speaking, this was not allowed, but the temptation was too great) the sun shone so brightly on my right that it was impossible to look in that direction.

Yevgeni Khrunov
USSR

Что больше всего поразило меня там, это – тишина. Немыслимая тишина, какой никогда не бывает на Земле, такая глубокая и полная, что начинаешь слышать собственное тело: как борется сердце, пульсируют сосуды, кажется, даже слышен шороха двигающих мускулов. А в небе звезд было больше, чем я мог представить себе. Абсолютно черное небо слегка подсвечивалось солнечным сиянием...

...Земля была такая маленькая, голубая и трогательно одинокая – наш дом, который нужно свято беречь... Она была идеально круглой. Я думаю, что я не понимал по-настоящему слова «круглый» до тех пор, пока не увидел Землю из космоса.

Алексей Леонов
СССР

What struck me most was the silence. It was a great silence, unlike any I have encountered on Earth, so vast and deep that I began to hear my own body: my heart beating, my blood vessels pulsing, even the rustle of my muscles moving over each other seemed audible. There were more stars in the sky than I had expected. The sky was deep black, yet at the same time bright with sunlight.

The Earth was small, light blue, and so touchingly alone, our home that must be defended like a holy relic. The Earth was absolutely round. I believe I never knew what the word round meant until I saw Earth from space.

Aleksei Leonov
USSR

Не торопясь, я вылез из люка, слегка
оттолкнулся от него и начал все
больше и больше удалятся от
космического корабля. В результате
этого , наш космический дом начал
медленно поворачиваться на моих
глазах.

Алексей Леонэв
СССР

I climbed out of the hatch unhurriedly
and gently pushed myself away from
it, moving farther and farther away
from the ship. This slight effort made
our spacecraft begin to turn slowly
before my eyes.

Aleksei Leonov
USSR

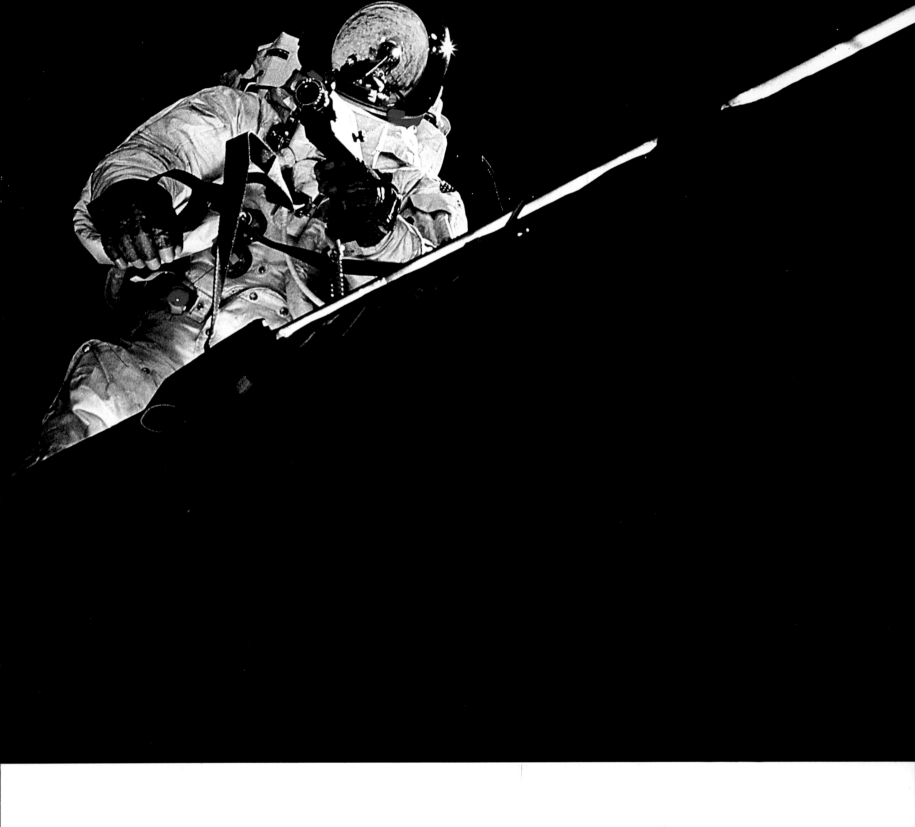

Я отправлялся в неизвестность, и ни один человек на земле не мог сказать мне, с чем мне предстоит встретиться там. Нет учебников. Это делается впервые. Но я твердо знал, что это нужно сделать. Ясно, что я должен быть очень осторожным.

— Спокойно, спокойно, — говорил я себе – не торопись.

Алексей Леонов
СССР

I set out into the unknown and nobody on Earth could tell me what I would encounter. There were no textbooks. This is the first time ever. But I knew for certain that it had to be done. It was clear that I had to be very careful. Take it easy, take it easy, I told myself, do not move too quickly.

Aleksei Leonov
USSR

Люк медленно открывался. Мы
стояли с Евгением, прижавшись друг
к другу, чтобы не мешать движению
крышки. В расширяющуюся щель
врывались солнечные лучи, которые
воспламеняли на противоположной
стенке отсека ослепительно яркий
зайчик. Мы смотрели в сторону
Земли. На темном фоне космоса она
казалась далекой и безжизненной.

Алексей Елисеев
СССР

The hatch opened slowly and Yevgeni
and I stood side by side so as not to
interfere with the motion of the lid.
Sunlight burst through the widening
gap and flamed against the opposite
wall of the compartment in a blinding
ray. We looked in the direction of
Earth. Against the black backdrop of
space it seemed distant and lifeless.

Aleksei Yeliseyev
USSR

Space Walk

The space suit always reminded me
of the activity that a four-year-old
youngster is put through when his
mother or father dresses him in a very
heavy snowsuit. Your mother doesn't
bundle you up but your shipmates do.
They put you in the space suit,
oftentimes with a pat on the head and
a butter cookie for good luck. Then
they put the helmet on the top, snap it
into place, and from that moment on
you float in the suit. You don't stand
on the boots. From time to time your
toes will touch the boots, but your
head will bob up against the helmet.
You are floating in this cocoon, and
you float out through the hatch. Using
the controls, you can maneuver
yourself away from the mother ship.
You are orbiting the Earth as surely as
the Moon orbits the Earth, and you are
yourself a satellite.

Joseph Allen
USA

Looking outward to the blackness of space, sprinkled with the glory of a universe of lights, I saw majesty – but no welcome. Below was a welcoming planet. There, contained in the thin, moving, incredibly fragile shell of the biosphere is everything that is dear to you, all the human drama and comedy. That's where life is; that's where all the good stuff is.

Loren Acton
USA

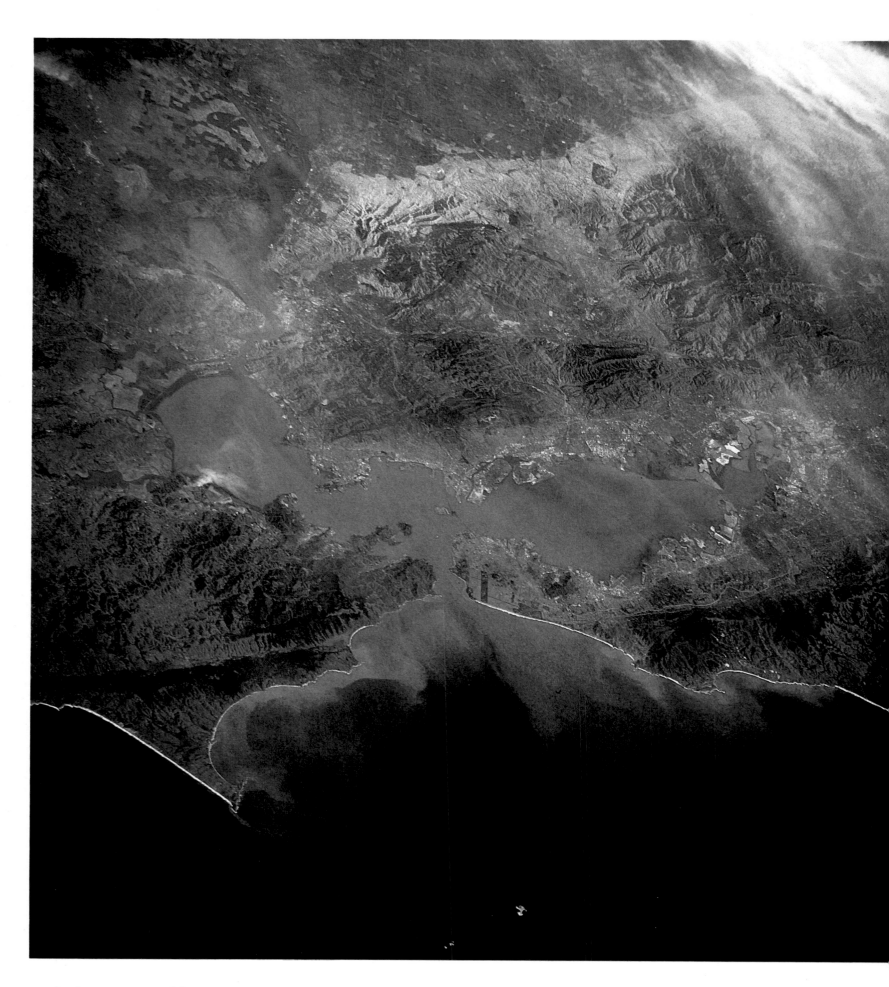

20   San Francisco Bay, California, USA

18   Laguna Verde in the Andes Mountains

There is a clarity, a brilliance to space
that simply doesn't exist on Earth,
even on a cloudless summer's day in
the Rockies, and nowhere else can you
realize so fully the majesty of our
Earth and be so awed at the thought
that it's only one of untold thousands
of planets.

Gus Grissom
USA

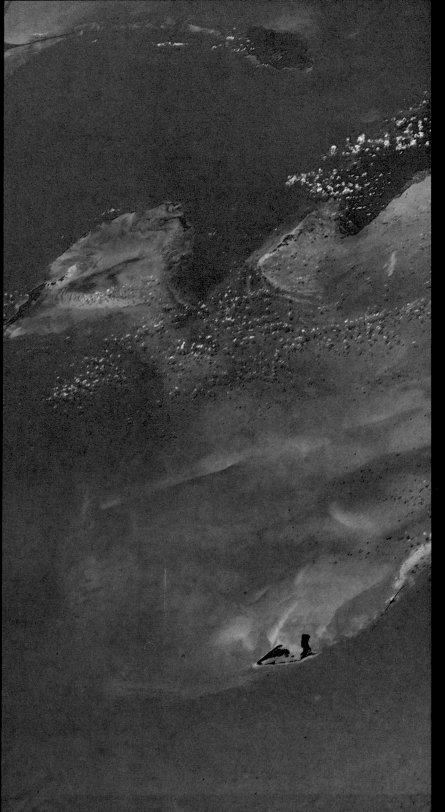

My first view – a panorama of brilliant deep blue ocean, shot with shades of green and gray and white – was of atolls and clouds. Close to the window I could see that this Pacific scene in motion was rimmed by the great curved limb of the Earth. It had a thin halo of blue held close, and beyond, black space. I held my breath, but something was missing – I felt strangely unfulfilled. Here was a tremendous visual spectacle, but viewed in silence. There was no grand musical accompaniment; no triumphant, inspired sonata or symphony. Each one of us must write the music of this sphere for ourselves.

Charles Walker
USA

7   Bahama Islands

The white, twisted clouds and the endless shades of blue in the ocean make the hum of the spacecraft systems, the radio chatter, even your own breathing disappear. There is no wind or cold or smell to tell you that you are connected to Earth. You have an almost dispassionate platform – remote, Olympian – and yet so moving that you can hardly believe how emotionally attached you are to those rough patterns shifting steadily below.

Thomas Stafford
USA

15 Bahama Islands

Испытываешь странное, будто
во сне, чувство полного физического
раскрепощения: раскинул руки, ноги
– и паришь. Однако надо быстро
научиться избегать твердых
предметов, стенок корабля. Помнишь
долго после исчезновения синяка, что
твоя масса всегда останется при тебе,
даже в другой Галактике.

Валерий Кубасов
СССР

You experience a strange dreamlike
sensation of freedom. You can spread
out your arms and legs as if soaring in
the clouds. You quickly learn to avoid
solid objects or the walls of the
station, though. Rubbing your bruise,
you remember that your mass remains
with you, even in another galaxy.

Valeri Kubasov
USSR

14   Coastal ranges, Canada and Alaska

В космосе все иначе. Спишь на
потолке, а за считанные минуты
пересекаешь целые континенты.
Нажимаешь на ключ, чтобы
отвернуть гайку, а сам вращаешься.

Анатолий Березовой
СССР

In space everything is different, you
sleep on the ceiling and you cross
whole continents in minutes. Push on
a wrench to loosen a nut and you find
yourself rotating.

Anatoli Berezovoy
USSR

I used to have dreams when I was a kid
that I'd go running down the street and
jump up in the air and go flying and
just fly through the air all by myself.
That's what weightlessness is like.

Robert Gibson
USA

We orbit and float in our space gondola
and watch the oceans and islands and
green hills of the continents pass by
at five miles per second. We move
silently and effortlessly past the
ground. I want to say "over the
ground" as I write this, but remember
that in space your sense of up or down
is completely gone and my description
must reflect this fact. In addition, the
breathtaking speed of the ship is in odd
and confusing contrast to the feel of
perpetually floating within the
spaceship. You do not sit before the
window to view the passing scene, but
rather you float there and look out on
the scene, certainly not down upon it.
Are you speeding past oceans and
continents, or are you just hovering
and watching them move beside you?

Joseph Allen
USA

Nieważkość pojawiła się nagle.
Wzniosłem się jakgdybym był
wewnątrz bańki mydlanej. Jak
niemowle w łonie matki, w moim
statku przestrzennym jestem ciągle
dzieckiem Matki Ziemi.

Mirosław Hermaszewski
Polska

Weightlessness comes on abruptly. I
soared as if I were inside a soap bubble.
Like an infant in the womb of my
spacecraft, still a child of my Mother
Earth.

Miroslav Hermaszewski
Poland

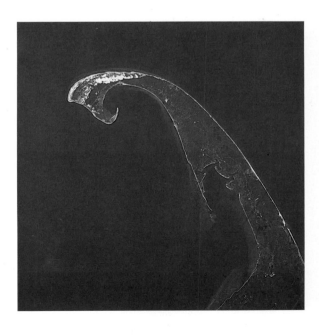

12   Cape Cod, Massachusetts, USA

11 Typhoon "Odessa" near Hawaii

Suddenly I saw a meteor go by underneath me. A moment later I found myself thinking, That can't be a meteor. Meteors burn up in the atmosphere above us; this was below us. Then, of course, the realization hit me.

Jeffrey Hoffman
USA

10   Tropical storm "Xina" over the Pacific

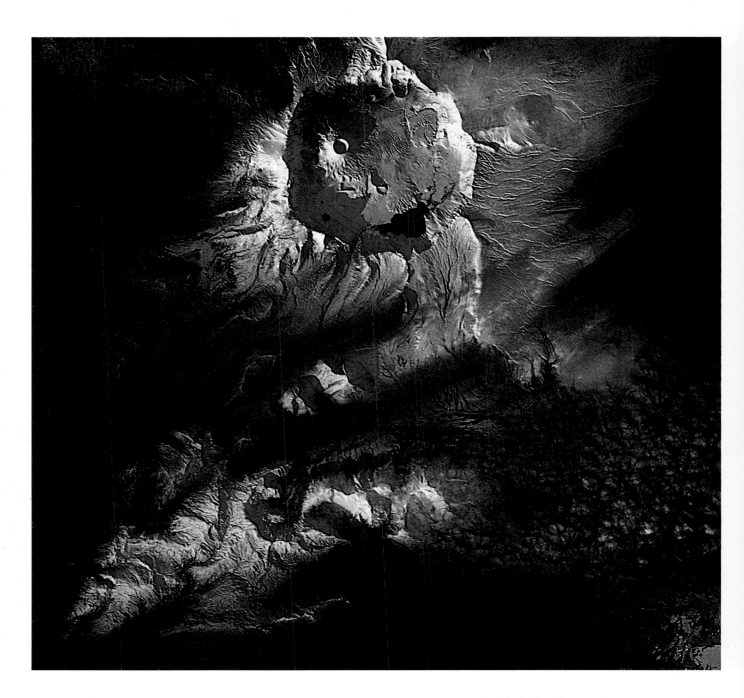

8 Aniakchak Volcano, Aleutian Range, Alaska

9 Klyuchevskaya Volcano, Siberia, USSR

When the engine shut down, I
unbuckled myself from my seat and I
was floating. I knew we were in orbit.
We had to do an orbital-maneuvering-
system (OMS) burn to get into a higher
orbit. But before we even did the burn
I floated upstairs – from middeck,
below the flight deck – to look over
the guys' shoulders. I looked out the
window and couldn't believe it. The
sun was streaming in, and you could
look right down at the Atlantic Ocean.
I looked at the three of them doing the
countdown for the OMS burn and I
thought, How in the world can you do
that? Look outside!

Joseph Allen
USA

7   Mozambique Channel, Madagascar

Моторын хуй нь мултарчихаад ракет мань бас л харанхуйд нисч байв. Анхамдаа од харагдаж байснаа, нэг зурвас хех ягаам огторгуйн туяа харагдаад, хелог онгоцны дотор мар туслаа.

Жугдэрдэмидийн Гуррагча
Монгол

The rocket went farther into the dark the moment it separated from the cowling. At first we saw stars, and then the sun entered the spacecraft, casting threads of the raspberry-colored cosmic dawn.

Zhugderdemidiyn Gurragcha
Mongolia

As soon as you get to an altitude and the vehicle starts to pitch over and accelerate downrange, then you start to be able to see the horizon. Once you get to some altitude – and I don't know what the number is – I would guess somewhere around 80 to 90,000 feet – you start to see that the sky is turning darker than we see it when we look outside, and the horizon begins to have a much sharper definition. It's not an Earth horizon but an atmosphere horizon, too.

Ken Mattingly
USA

6   Dawn over the Pacific Ocean

Мне стало казаться, что я покинул
планету навсегда. И что нет силы,
которая может вернуть меня обратно.

Валерий Рюмин
СССР

It seems I am leaving the planet
forever. And there is no power that
could bring me back.

Valeri Ryumin
USSR

The *Atlas* is an eerie thing to sit on top
of when the gantry is gone. I could
hear the sound of pipes whining below
me as the liquid oxygen flowed into
the tanks and a vibrant hissing noise
as they were supercooled by the *Lox*.
The *Atlas* is so tall and limber that it
sways slightly in heavy gusts of wind,
and in fact, I could set the whole
structure to rocking a bit by moving
back and forth in the couch. I could
look at the whole sky now with its
fast-growing patches of blue. Through
a mirror mounted near the window I
could see the blockhouse and across
the Cape. Through the periscope I
looked east at the Atlantic along the
track I would follow.

John Glenn, Jr.
USA

I heard the word ignition and I sensed,
felt, and heard all the tremendous
power that was being released
underneath that rocket beginning to
lift me off the Earth. It was a moment
of just supreme elation, complete
release of tensions. There were tears
coming down my face that morning.

James Irwin
USA

3   *Soyuz T-13* on the launch pad

Then they closed the hatch; it went clang like a dungeon door.

James Irwin
USA

They started to put the hatch on. This is the moment when things begin to come home to you. Up to this point people are reaching into the capsule and working all around you and there's no real feeling of being on your own. Suddenly, as people begin to pat you on the shoulder, wink at you, shake your hand and wave good-bye, it changes.

John Glenn, Jr.
USA

As we drove out, there was plenty of
time to reflect on your life: Did you
have your life in order; where are you
going today or where are you hoping to
go today? By this time it was light:
The sun was up and I could see that it
was going to be a beautiful morning,
no clouds that might interfere with
our launch; I could imagine that the
birds were singing on the outside, but
on the inside it was so quiet. We
weren't talking, we looked at each
other a little bit and smiled, that smug
look: we knew where we were going.
We were quiet because there were so
many thoughts racing through our
minds. When we got out to the pad, we
got out of the van and walked over to
the elevator. We walked slower that
morning as I recall; we were looking
around a lot; we didn't want to miss
anything, because for all we knew it
might be our last time to see things on
Earth. What caught our attention was
the spacecraft, the rocket, the *Saturn 5*,
with our little space module perched
way on top, a beautiful sight, just
gleaming white in that early morning
light. We viewed it in a little more
personal terms, and we just wondered,
you know, will it work? Will it really
take us to the moon and bring us back?

James Irwin
USA

Мы шли по плоской, как стол, казахской степи. Впереди, на фоне удивительно четкого горизонта, серебряной стрелой вонзается в небо стоящая на старте ракета. Она дышит, и клубы белого пара обтекают ее стройное тело, укрытое подвенечным платьем инея. Она ждет нас, а мы идем к ней. Дублеры все время рядом. Шутят, но в глазах видна грусть. Готовились к полету, но на работу в космос уходят одни, а другие остаются. Им снова ждать своего часа.

Александр Волков
СССР

We walked across the table-flat Kazakh steppe. In front of us and against the background of an incredibly distinct horizon, the rocket stood on the launch pad pointing into the sky like a silver arrow. The rocket was breathing, a cloud of white vapor flowing from her slender shape, which was covered in a bridal gown of frosting. She was awaiting us and we walked toward her. Our stand-ins were close-at-hand all the time. They were joking, but you could see sadness in their eyes. They were prepared for the mission, but only a few go to work into space, the rest must stay behind. They would have to wait again for their turns.

Aleksandr Volkov
USSR

I: Ai dormit în noaptea dinaintea zborului?
R: Da, am dormit.
I: Adevărat, fără să iei somnifere?
R: Fără somnifere, dar am avut o noapte îngrozitoare.

Dumitru Prunariu
România

Q: The night before you went up — did you sleep?
A: Yes, I did sleep.
Q: Really, without sleeping pills?
A: Without pills, but I had a terrible night.

Dumitru Prunariu
Romania

Outward

Виталик ласкается, целует меня, – чувствует, милый мальчик, что отец улетает надолго.

Перед уходом из дома мы сели за стол на кухне и по традиции поставили хлеб, соль и воду.

Когда отъезжали от дома, я посмотрел на балкон и увидел, что мама вытирает слезы. Я помахал ей рукой, но она меня не видела.

Утро прекрасное. Ясное небо, солнце, свежесть. Только на душе как-то неспокойно.

Валентин Лебедев
СССР

Vitalik caresses me and kisses me; he senses, dear boy, that his father is flying off for a long time.

Before leaving the house we sat at the kitchen table. By tradition, there was bread, salt, and water.

When we drove away from the house I looked to the balcony and saw Mother wiping away the tears. I waved to her; she did not see me.

A fine morning. A clear sky, sunny, fresh. Only in my soul is there something unquiet.

Valentin Lebedev
USSR

Ich hatte mir gewünscht, daß die Menschen mich nach meiner Rückkehr gefragt hätten, wie es mir dort ergangen ist. Wie ich mit der glitzernen Schwärze der Welt fertig geworden bin, und wie ich mich als Stern, der die Erde umkreist, fühlte.

Reinhard Furrer
Bundesrepublik Deutschland

I would have wished that after my return people had asked me how it was out there. How I coped with the glistening blackness of the world and how I felt being a star that circled the Earth.

Reinhard Furrer
Federal Republic of Germany

A crescent Earth

The Home Planet

ments from the astronauts and cosmonauts is that the photos, as beautiful as they are, never quite capture the ineffable quality of viewing the Earth from space. The colors aren't as bright, there are not as many shades, you cannot see in three dimensions, and of course the all-important backdrop is lacking.

The Apollo astronauts say that from the moon the Earth looks like a small, delicate, blue and white marble. Stick out your thumb and you can blot out everything that has any meaning to you on that fragile globe, small enough to crumble between your fingers. The first astronaut I interviewed was Jim Irwin. "Think what it will be like," he said, "for those men who go to Mars and see our Earth shrink to the size of a star, just a bright blue star in the heavens. Think what those men will feel for it then."

There is no way to be sure or to really know what these people will experience, but one thing is clear: Space offers us a chance to see our world with new eyes, a perspective that may have great significance for the planet for all of the future.

Something has happened to me in the making of this book, searching through the tens of thousands of images, interviewing astronauts and cosmonauts, and reading countless impressions and recollections about space. The context of my reality has broadened. I will experience it most directly tonight when I turn off the light and close the door of my studio and trace my path home in the darkness. I will be aware that I am walking on a round planet hurtling around the sun at 62,000 miles an hour, turning at 1,000 miles an hour at the Equator, producing day and night. I will be aware of our Sun as the center of a Solar System that is moving around the Galaxy at more than 500,000 miles an hour, and of the whole Galaxy itself hurtling in a direction unknown to me at an unimaginable speed through an ever-expanding universe populated with billions of other galaxies stretching to eternity. I think this sense of wonder at our universe and the strangeness of our lives within our tiny part of it is important to our sense of ourselves and perhaps to our very survival. I hope this book will help you see, and will add to your appreciation of, the great beauty, the incredible wonder, and the unfathomable mystery of all this as it unfolds in the eternal moment.

As I write, five men float in their space station 400 miles starward, gliding at 17,000 miles an hour through a cold and empty blackness. Physics tells us that the pull of gravity upon their bodies and their ship decreases with the square of their distance from Earth. But based on my experience in preparing this book, I can only assume that Earth's pull on their hearts grows with every mile.

Many astronauts and cosmonauts, upon returning from their missions, report changes that are powerful and life-transforming, whole-life changes they attribute to the simple experience of looking back at our home planet from the remoteness of space.

I too have seen the distant Earth silently spinning, but only in daydreams. Yet this vision has always brought me a deep sense of comfort and peace. I wondered, then, if my imaginings could be so moving, and the authentic, visual experience so much more so, why not find a way for all of us, the millions who will never be astronauts or cosmonauts, to have some taste of it? Why not gather the most beautiful photographs ever taken of Earth from space and combine them with the thoughts and feelings of those who have seen the sights firsthand?

From a visit to the Air and Space Museum in Washington, D.C., I knew that extraordinary images of Earth existed, and I assumed there had to be more in the archives. Which left only one problem: getting in touch with astronauts and cosmonauts from all over the world and gathering their innermost thoughts.

I was most fortunate to be able to join forces with the Association of Space Explorers, an international organization of individuals who have orbited Earth at least once. Their collaboration and support were essential in allowing me to gather the material I needed. I was surprised, and delighted, by the wide variety of observations that never would have occurred to me in my most vivid daydreams. But most rewarding was the intimate and sometimes spiritual story that emerged through these recollections, revealing that, whether on a week-long trip to the moon or nearly a year in the confinement of a space station, we take our humanness with us into space, and there it seems to be enhanced.

The search for photos began by my looking at every medium- and large-format handheld image in the NASA archives. In time, I would examine every available image from the Soviet archives as well, along with images from the most advanced mapping cameras, including the 8″ x 19″ large-format camera (LFC) flown on the shuttle. Editing the 2,500 or so photographs chosen from the tens of thousands viewed was a painfully slow and difficult process. If only we had had room for 600 photos it might have been easier.

While I endeavored to have the photos cover a wide range of colors and textures, landforms and weather patterns, there was no effort to choose photos by their geographic location. In fact, it was only after the photographs were placed in the book that I found out what parts of the world many of them represented. The overriding criterion for the selection of photographs was their beauty and their ability to capture the many moods of the Earth. I was also interested in making sure we included photos that captured the Earth's diversity – a broad range of geological formations, oceans, clouds, mountains, deserts, volcanoes, and islands. In the back of the book you will find a detailed description of each of these images.

The photos in this book represent only a very narrow and selective glimpse of our planet. You should not be deceived into believing you have "seen the Earth" by viewing the book. In fact, one of the most frequent com-

Introduction

Kevin W. Kelley

cosmic abyss that engages your attention, but the spectacle of our small planet haloed in blue. Suddenly, you get a feeling you've never had before, that you're an inhabitant of Earth.

Your whole body feels the roaring power of the rocket, and you know it obeys the navigation commands perfectly, yet unconsciously you keep returning to the thought that only very thin walls separate you from the deathly cold and incomprehensible emptiness of space, which can extinguish life instantly and pitilessly. You look down upon the Earth with mixed feelings of delight and adoration. It's not just that beneath you is your home, the whole of your world, but that the power and strength you have been given was generated by humankind's intelligence and ability, and maybe they are unique in this endless universe.

I believe that these or similar feelings and thoughts were the stimulus for the creation of a special brotherhood of inhabitants of Earth. Those who have been in space realize that, in spite of the complete disparity between them, they are one in an important way, namely, an acute feeling of being an inhabitant of Earth, a feeling of a personal responsibility to preserve the only planet we have. They realize that any predicament, disagreement, or obstacle can be overcome.

In 1981, several people including Mike Murphy, Jim Hickman, Georgi Arbatov, and Andrei Kokoshin thought of and began to pursue the brilliant idea of forming the Association of Space Explorers. The first working meeting, which took place after lengthy negotiations, occurred in 1983 in Pushchino near Moscow, with Russell Schweickart, Mike Collins, Ed Mitchell, Aleksei Yeliseyev, Aleksei Leonov, Vitali Sevastyanov, and Valeri Kubasov taking part. It became

apparent that something could and should be done together. Our major anxiety and personal responsibility was to protect and conserve the Earth's environment. We decided that, if the efforts we were putting into the foundation of our Association were to have results, then the theme of our first congress, "The Planet Our Home," should be our main one. We also decided to give an award to the person whose life is an example of consideration to all living things and to the riches of the natural world. We agreed unanimously that the prize should go to Jacques-Yves Cousteau, who said to us, "You have helped us understand the stars. You have changed our ideas about humanity, space, and the unknown, and that is important for future happiness."

We have since met again, had discussions, written long letters to each other, and held extended telephone conversations, because we have found that it is not so easy to agree on the variety of details. We all come from different countries, and on Earth the distances that separate them seem so enormous; it seems that only from outer space does our blue planet resemble a touchingly small sphere. But all the differences and difficulties can be overcome, and the right words can be found, when we are united by a common important goal – a goal that is really so simple – to make our conviction and knowledge more understandable to every dweller on Earth, and to convey it to them more quickly. We hope that everyone will come to share our particular cosmic perception of the world and our desire to unite all the peoples of the Earth in the task of safeguarding our common and only, fragile and beautiful, home.

Cosmonauts don't say much, especially when we're on a mission. Usually some five to seven seconds are enough for us to express the most complicated thought. Or so I thought; but then a while ago I was asked to listen to recordings of the communications between mission control and the cosmonauts in space. To tell you the truth, I was amazed. Within seconds of attaining Earth orbit, every cosmonaut, without exception, be they a dry, reserved flight engineer or a more emotional pilot, uttered the same sort of confused expression of delight and wonder.

Curious, I analyzed the initial conversations for a variety of missions and discovered an interesting pattern. It didn't matter whether the cosmonaut was on a one-man mission in the first Vostoks or part of a large crew on a mission in a modern Soyuz, no one has been able to restrain his heartfelt wonder at the sight of the enthralling panorama of the Earth. The emotional outbursts lasted forty-two seconds on average.

Anyone who has seen the Earth from space knows that it is an incomparable sight. It's not just that the planet is piercingly beautiful when viewed at a distance; something about the unexpectedness of the sight, its incompatibility with anything we have ever experienced on Earth, or known, or practiced, elicits a deep emotional response.

I was entranced when I saw the whole American continent from the Pacific to the Atlantic illuminated for a fleeting instant by an incredible sunbeam; the mirror of the Amazon basin, with its swamps and backwaters, like the bewitching eye of the continent, flashing up a friendly wink: Earth's greeting to space, the stars, a speck of dust – our space capsule.

I remember something from my first flight with Vasili Lazarev. I was hanging in the orbital module preparing to engage the nine-lens miniaturized camera to photograph a corridor across the USSR from the Black Sea to the Pamirs, to verify whether it was worth taking spectrozonal photographs of the Earth from orbit. I should tell you that, when the porthole is pointing straight down at the ground, you can't see the horizon, and so you get the impression that you are watching a map gliding by beneath you, a map just like the ones at school with mountains and rivers and landmasses and oceans all perfectly inscribed. Unconsciously, you look for the lines that are usual on such maps, the parallels and meridians; it is strange not to see the markings on the living map. However, the colors that once were chosen for physical maps are almost completely true to life: the reddish tinge of the Sahara, the yellow of the deserts of Central Asia, and the sapphire blue of the oceans. Only from space can you see that our planet should not be called Earth, but rather Water, with specklike islands of dryness on which people, animals, and birds surprisingly find a place to live.

The artistic genius who painted our planet worked from a fantastic assortment on His palette and with unusually pure colors. We flew across the Crimea. It was autumn and corn was ripening in the Kuban area, the delicate yellows of the grain iridescent in a hundred shades. Fifteen minutes later we saw the soft green of the forest in the Taiga, then the dark brown of the Himalayas, and once again the long stretch across the lovely sapphire ocean.

Anyone who has been in space knows that the impatiently awaited unearthliness quickly loses its charm. It is not the boring uniform blackness of the

Preface

Oleg Makarov

feelings and impressions of our home planet in unique ways. And yet it is the golden thread that runs through all these expressions of individual experience that is the magic of life. We spend a great deal of time identifying and emphasizing the differences between things in our professional roles, including ourselves. And yet it is our common human experience, our shared fear, hope, joy, and love, that link us as human beings beyond all differences.

It is this shared personal impression of our home planet that has brought many of us together as the Association of Space Explorers. We hope you too will experience this new connection between us humans and our home planet as you read this book. It is the golden thread that connects us all and which I hope you will ponder long after the beauty of the specific images fades in your memory. It is what I ponder now, and what I will marvel over the rest of my life.

experienced those uniquely human qualities: awe, curiosity, wonder, joy, amazement. It is these shared human experiences, the physical and the emotional, that link those of us who have flown in space about this planet. Long after the mission is over it is the reflection on the personal interaction with the experience that stays alive.

For me, having spent ten days in weightlessness, orbiting our beautiful home planet, fascinated by the 17,000 miles of spectacle passing below each hour, the overwhelming experience was that of a new relationship. The experience was not intellectual. The knowledge I had when I returned to Earth's surface was virtually the same knowledge I had taken with me when I went into space. Yes, I conducted scientific experiments that added new knowledge to our understanding of the Earth and the near-space in which it spins. But those specific extensions of technical details I did not come to know about until the data I helped to collect was analyzed and reported. What took no analysis, however, no microscopic examination, no laborious processing, was the overwhelming beauty . . . the stark contrast between bright colorful home and stark black infinity . . . the unavoidable and awesome personal relationship, suddenly realized, with all life on this amazing planet . . . Earth, our home.

For me, this experience was a dramatically enlarged version of looking out over the hills and valleys spreading into the distance after having climbed a high mountain peak. It is not simply the view. It is smelling the perfume of hot crushed pine needles along the way, catching glimpses of the Douglas squirrel scolding the intruder from his private fir tree, gazing in awe from the warm sleeping bag in the middle of the night at the countless stars arrayed just overhead. All these quiet personal connections are there filling the heart when finally you look out over the receding trees from the top of the mountain. It is, in fact, a visual embrace with all that life with which you are connected. So too with the space experience. For me, it was an embracing of the planet and all the life on it . . . and, like the squirrels and the pines, it hugged me back.

As I thought about it afterward, as I relived the experience time and again, I came to understand that this was not some sentimental recollection of past glory. Rather, I began to understand that it is the personal manifestation of a relationship which, in the absence of direct experience, we can know only intellectually. We all understand that the life systems of this planet are interrelated, that our human future depends on the well-being of the rain forest and the salt marsh. We know that human activity in the production of goods and services can damage and destroy the environment on which we and our children depend. We know all these things intellectually. Yet we feel related to all people when we see pictures of mothers and children, tears of sorrow and joy, laughter, music, and dance. And we fear together the misuse of the power we have now at our collective fingertips through our amazing technology. What the experience of seeing this amazing planet from space does is to take it beyond the intellectual and into the personal.

I suspect that each of us who has had this experience of circling the planet again and again would express it differently. We are all different people. We come from different cultures. And even within the same culture we have vastly different experiences and origins. It is therefore not unexpected that we would express our

When people in the future look back on these early days of space exploration, they will create many images of what it must have been like to be among the first people who lived and traveled beyond Earth's atmosphere. There will be those who glamorize the experience to the point of painful embarrassment for those few who experienced it. Others will see the endeavor in terms of geopolitical struggle, or the scientific quest for understanding our cosmic environs and our place in it. Still others will emphasize the inexorable pressure of technology to extend the capabilities of humankind, to understand the planet on which we live, and the commercial benefits to be reaped by those who "get there first."

In fact, for those of us who lived this unique experience, the incredible excitement of being selected as an astronaut or cosmonaut was soon displaced by the daily repetition of very hard work. Thousands of hours were spent in simulations, training sessions, mock-ups, stowage reviews, checklist reviews, mission rules meetings, and so forth. Thousands of people were involved in the process of getting a spacecraft into space, each with his or her large or small part, virtually none with a *truly* comprehensive view of the entire process. Even we who flew the missions didn't know, couldn't know, everything about the launch vehicles or spacecraft. At times the complexity and staggering detail of the endeavor was so overwhelming that it was impossible to imagine it would all culminate in actually blasting your soft pink body off the planet atop a towering pillar of flame.

Yet that day came for each of us who writes in this book. None of us will forget that morning (or afternoon) of launch. How ordinary it seemed; like any other in so many ways. And yet how different! You go around in two pieces, one piece of you doing all the ordinary things that need to be done, and the other, watching with a sense of unreality, disconnected.

Until getting into the spacecraft. Then it becomes real. All the preparation, all the simulations, all the reviews are now behind. The day has finally come. It's countdown time. You hear and feel the vibration and the power of those engines. And then you start slowly up, not with a rush, but majestically, almost defiantly. The ingenuity and determination of technology in breaking another barrier – gravity – are manifest, with you as the ultimate witness.

But you don't have time to think about it. You are too busy. Whether you have a short stay or long, whether orbiting the Earth or heading moonward, you have little or no time for idle chatter, sightseeing, or even thoughts. Good housekeeping in a spacecraft is not merely desirable, it is mandatory. Weightlessness demands that everything be put in its proper place, usually in a container or locker. Hours are spent in packing and unpacking, for anything not put away is guaranteed to float disruptively into the path of whatever critical task is next at hand. Fans whir, pumps hum, air hisses, the radio buzzes with the special language of numbers and tailored acronyms.

Yet we are human. We are not machines. We catch a glimpse of a huge swirl of clouds out the window over the middle of the Pacific Ocean, or the boot of Italy jutting down into the Mediterranean, or the brilliant blue coral reefs of the Caribbean strutting their beauty before the stars. And in the moments that each of us took away from our scheduled sleep time, or while waiting for some experiment to complete a sequence, we

Preface

Russell L. Schweickart

Like most fathers, by clear star-studded skies I used to take each of my two little boys in my arms for a glimpse at infinity. The splendor of the unreachable silenced their chatterboxes for a few seconds. They raised their arms and closed their little fingers in a futile attempt to grasp one of the twinkling sparks that dot our dreams. The little fellows obeyed the command reported by Ovid: "God elevated man's forehead and ordered him to contemplate the stars."

None of us three was gifted with a space explorer's destiny, and I know that deep in our subconsciouses a regret is buried. But the three of us have dedicated our lives to explore another infinity, the sea. And each time we have met, discussed, or worked with space pioneers, we have felt like brothers.

My father was ninety-one when the Apollo program was deployed. He never believed that a human being would ever reach the moon. When Borman, Lovell, and Anders, in *Apollo 8*, spun ten times around the moon, he told me: "Wonderful. Extraordinary. But believe me, they will never *land* on the moon." Unfortunately, he died just before Armstrong made his historic first walk on the queen of our nights.

Since then, probes have been sent to most of the Sun's planets; space stations have been put in orbit to demonstrate that human beings can live and work in outer space; astronauts and cosmonauts have been able to leave their space stations, walk in space, and even make essential repairs. A program to land people on Mars is in preparation.

The acquired technology has immediately been aimed at practical and profitable applications: worldwide communications, global positioning systems for ships and aircraft, and remote sensing to better know our planet and monitor its resources and to trace migrations of whales, fish, and birds. Unfortunately, it is now almost monopolized by the military.

This fallout has proved to be surprisingly beneficial to knowledge, to science, and to the future of humanity. But it was unforeseeable when Gagarin was shot into orbit on board *Vostok I* in 1961. At the origin, the pioneers of the greatest adventure of all times were motivated by the drive to explore, by the pure spirit of conquest, by the lofty desire to open up new fields to human genius. When I listened to the space explorers of all countries recently gathered in a meeting at Paris, they were all friendly, exchanging such feelings.

From their exceptional journeys, they all came back with the revelation of beauty. Beauty of the black sky, beauty and variety of our planet, beauty of the Earth seen from the Moon, girdled by a scintillating belt of equatorial thunderstorms. They all emphasize that our planet is one, that borderlines are artificial, that humankind is one single community on board spaceship Earth. They all insist that this fragile gem is at our mercy and that we must all endeavor to protect it.

The meaning of space conquest is symbolized by the famous set of pictures taken from the moon, celebrating the birth of a global consciousness that will help build a peaceful future for humankind. That future is in the hands of those who dedicate their lives to explore Teilhard de Chardin's three infinities: the infinitely big, the infinitely small, and the infinitely complex. And from all the beauty they discover while crossing perpetually receding frontiers, they develop for nature and for humankind an infinite love.

Foreword

Jacques-Yves Cousteau

Dedicated to all the children of the world

Queen Anne Press

© Kevin W. Kelley 1988

First published in Great Britain in 1988 by
Queen Anne Press, a division of
Macdonald & Co (Publishers) Ltd
3rd Floor
Greater London House
Hampstead Road
London
NW1 7QX

A Pergamon Press plc company

Published by arrangement with Addison-Wesley Publishing Company,
Reading, Massachusetts, USA.

British Library Bibliographic Services Cataloguing in Publication

The Home Planet.
1. Earth. Surface features. Aerial photographs
I. Kelley, Kevin W.
525'.022'2

ISBN 0-356-15984-1

Frontispiece:

Earth seen from 23,000 miles away

The
Home
Planet

Conceived and edited by
Kevin W. Kelley
for the Association of
Space Explorers

This book was funded in part by the
Institute of Noetic Sciences

Original design concept by Carol Denison

Macdonald
Queen Anne Press

Addison-Wesley Publishing Company
Reading, Massachusetts   Menlo Park, California   New York
Don Mills, Ontario   Wokingham, England   Amsterdam   Bonn
Sydney   Singapore   Tokyo   Madrid   San Juan

Mir Publishers
Moscow

De ruimte is zo vlakbij: binnen de
acht minuten waren we er en in
twintig minuten waren we weer
terug.

Wubbo Ockels
Vrij Nederland

Space is so close: It took only eight
minutes to get there and twenty to
get back.

Wubbo Ockels
Netherlands

Странное чувство полного, какого-то торжественного покоя вдруг охватило меня, когда спускаемый аппарат качнулся и застыл. Была непогода. Я вдохнул запах земли – несказанно сладкий и хмельной. И ветер. Какое же это наслаждение – ветер после долгих дней в космосе.

Андриян Николаев
СССР

A strange feeling of complete, almost solemn contentment suddenly overcame me when the descent module landed, rocked, and stilled. The weather was foul, but I smelled Earth, unspeakably sweet and intoxicating. And wind. How utterly delightful; wind after long days in space.

Andriyan Nikolayev
USSR

127   The Saint Elias Range and the Gulf of Alaska

И вот я снова стою на Земле. Меня слегка покачивает. Вокруг, насколько хватает глаз, простирается серая, осенняя степь. Ни куста, ни деревца, только вертолет поисково-спасательной службы, словно уставшая большая птица, присел в сторонке. Но как я был рад этой, уже слегка припорошенной первым снежком, земле. Мне захотелось упасть на нее, обнять, прижаться к ней щекой!

Георгий Шонин
СССР

Once again I stand on Earth. I wobbled somewhat. All around, as far as the eye could see, stretched gray autumnal steppe. Not a shrub, not a tree in sight, only the helicopter of the search and rescue service sitting nearby, looking like some large tired bird. I was so happy to see the ground, already a little covered by the first fluffy snow. I wanted to fall into it, hug it, and press my cheek to it.

Georgi Shonin
USSR

28   Volcanic regions of northwest Argentina

Reflections

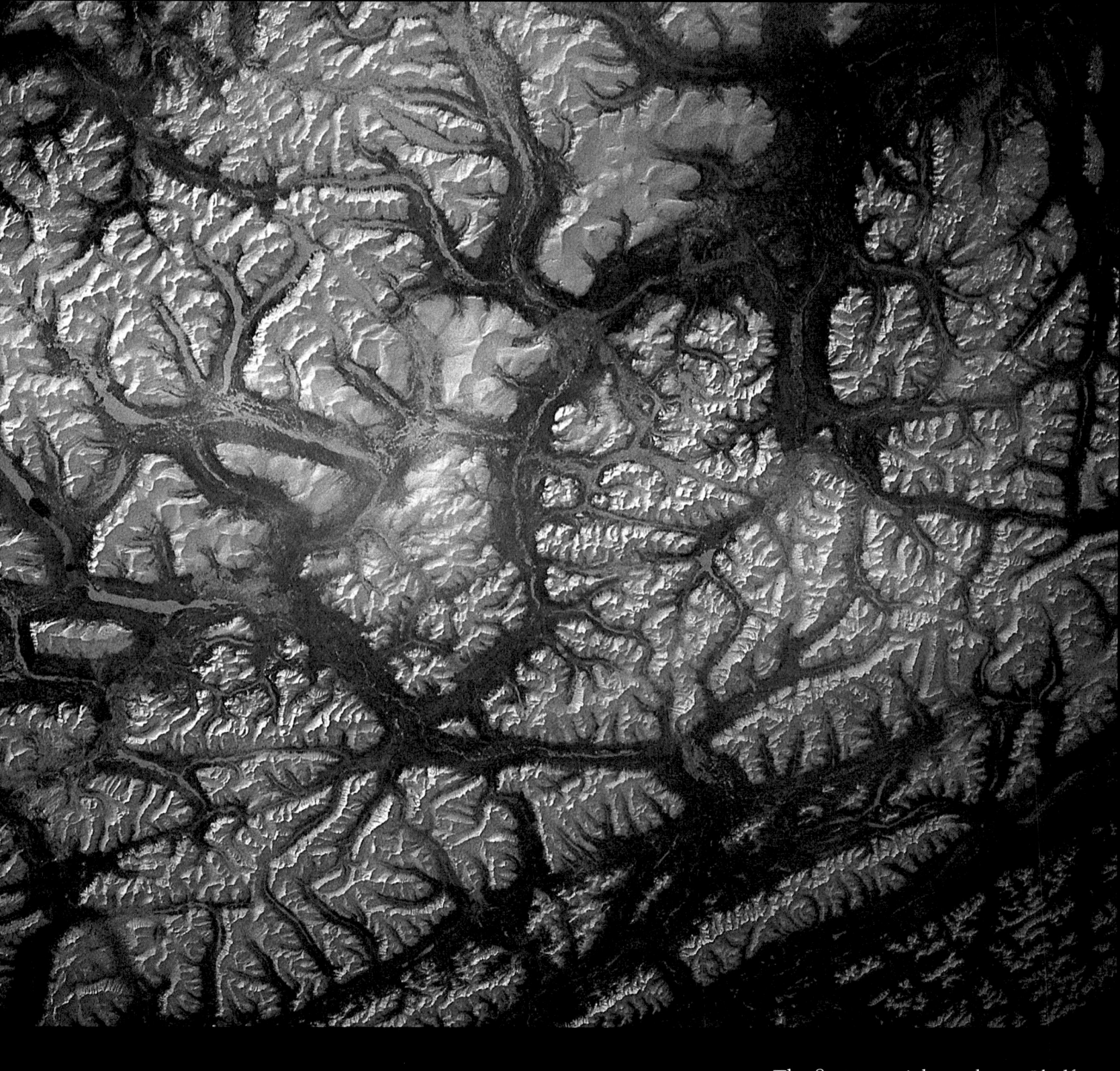

The first two nights at home I half woke up and reached back in the dark to feel for the structural struts of the spacecraft. They didn't feel right. As I gradually came out of deep sleep it would hit me: It's not metal – it's wood. And then I would realize that I wasn't holding a strut at all, but the head of the bed.

Thomas Stafford
USA

The peaks were the recognition that it is a harmonious, purposeful, creating universe. The valleys came in recognizing that humanity wasn't behaving in accordance with that knowledge.

Edgar Mitchell
USA

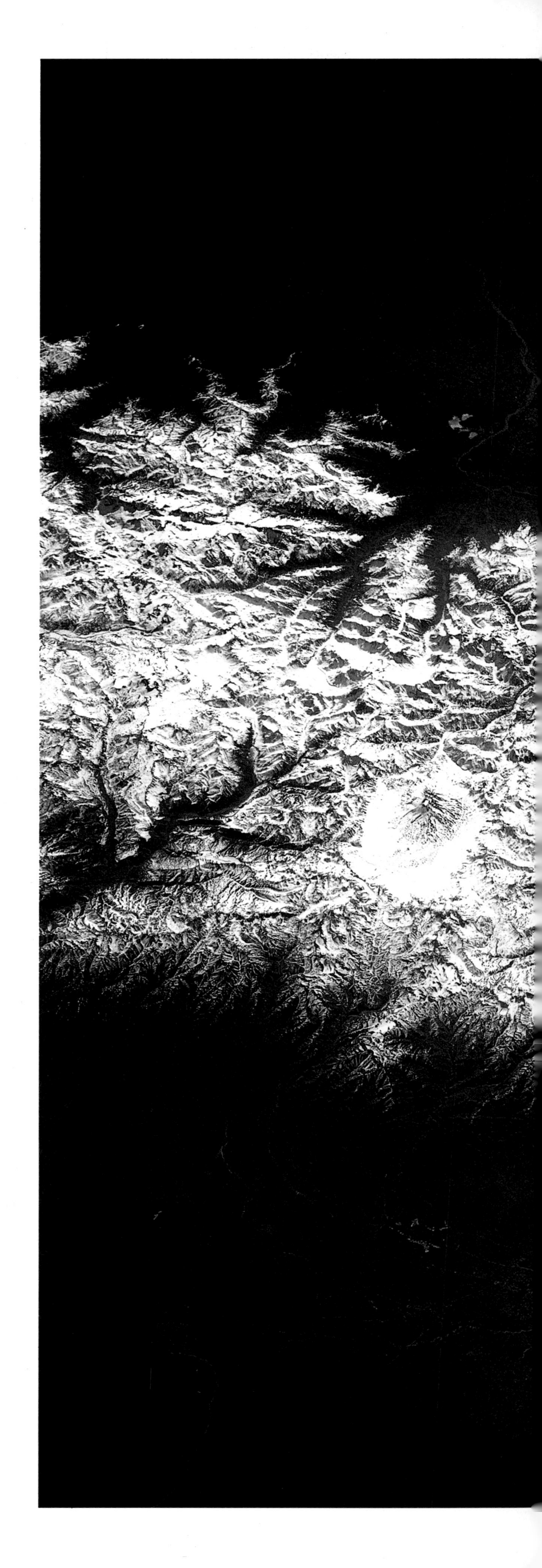

132   The Nile and the Nubian Desert, Sudan, Egypt

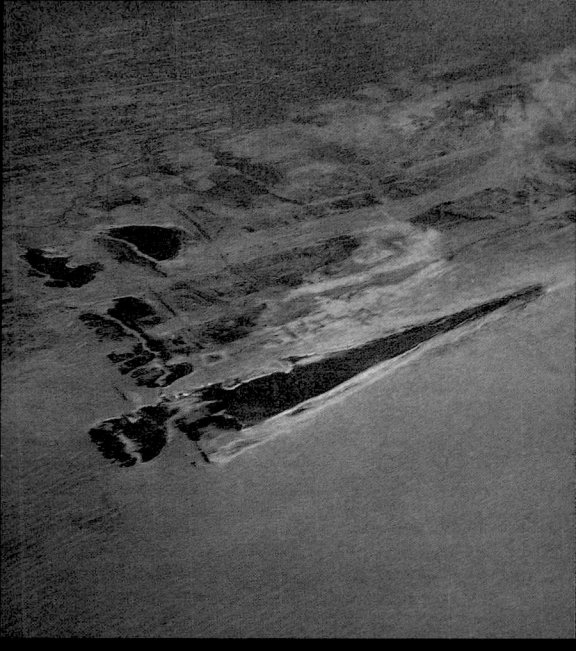

133   Lake Fatuibine, in Mali

When the history of our galaxy is
written, and I think it may already
have been, if the planet Earth gets
mentioned at all, it won't be because
its inhabitants visited their own
moon. The first step, like a newborn's
first cry, would be automatically
assumed. What will be worth
recording is what kind of civilization
we Earthlings created and whether or
not we ventured out to other parts of
the Galaxy. Were we wanderers?
Human history so far indicates we
are indeed.

It's human nature to stretch, to go,
to see, to understand. Exploration is
not a choice, really, it's an imperative.

Michael Collins
USA

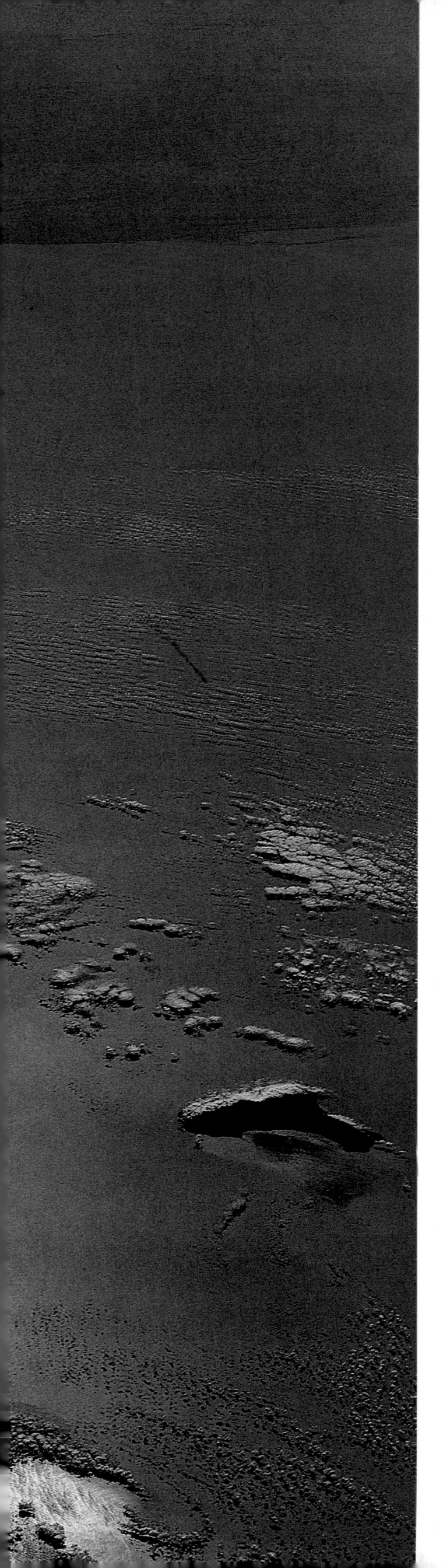

Ветры разносят по планете семена
жизни, рождая травы, леса. Мчатся
вечные ветры Вселенной. Что несут
они? Никто не знает. Но я уверен, что
природа создала нас, наделенных
разумом, чтобы подобно ее слугам —
ветрам мы несли жизнь в
необозримые, бесконечные просторы,
в ее неисчислимые миры Вселенной.
Добрый разум должен победить на
Земле, а потом и во всей Вселенной.

Юрий Глазков
СССР

The winds scatter across the planet
the seeds of life to bring forth the grass
and flowers and woods. The eternal
winds of the universe are rushing
along. What do they bring? No one
knows. But I am sure that Nature
has created us, endowed us with
intelligence, so that we, like her
servant the winds, can carry life into
the vast and limitless emptiness and
to its innumerable worlds. Reason
should win out on Earth and then in
the whole universe.

Yuri Glazkov
USSR

In space one has the inescapable
impression that here is a virgin area of
the universe in which civilized man,
for the first time, has the opportunity
to learn and grow without the
influence of ancient pressures. Like
the mind of a child, it is yet untainted
with acquired fears, hate, greed, or
prejudice.

John Glenn, Jr.
USA

136   Rio de Janeiro Bay, Brazil

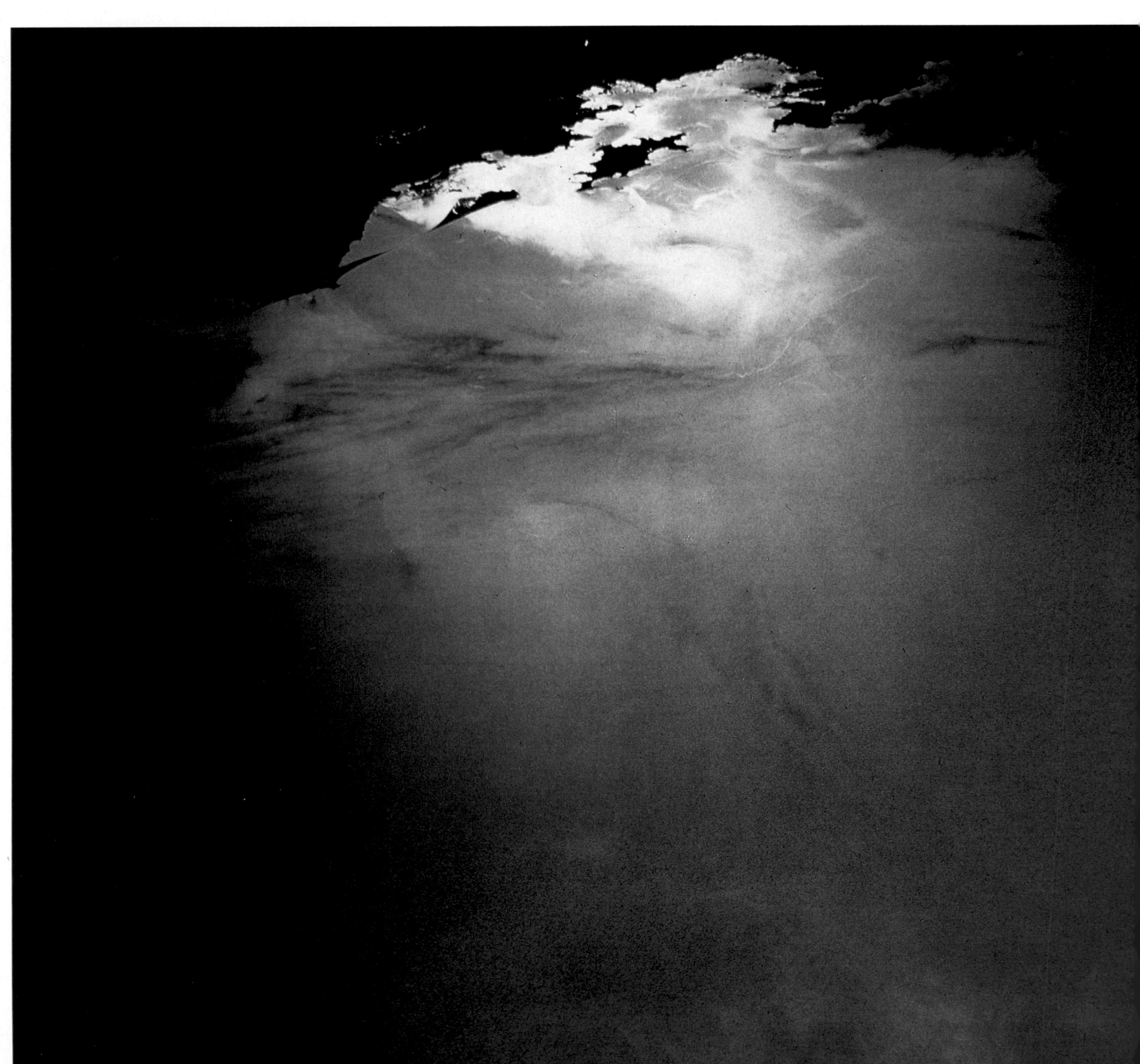

137   Cumulonimbus clouds over the Congo Basin, Zaire

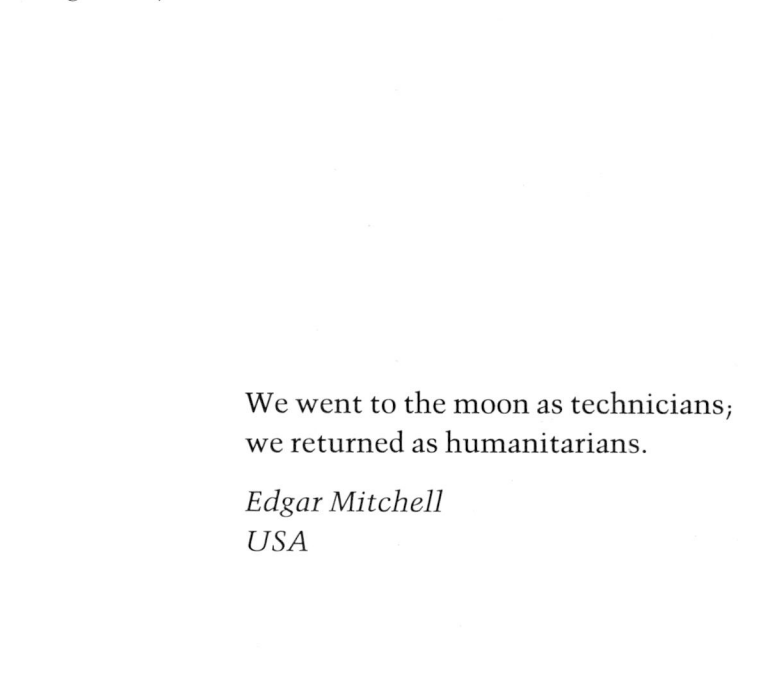

We went to the moon as technicians;
we returned as humanitarians.

Edgar Mitchell
USA

Instead of an intellectual search, there was suddenly a very deep gut feeling that something was different. It occurred when looking at Earth and seeing this blue-and-white planet floating there, and knowing it was orbiting the Sun, seeing that Sun, seeing it set in the background of the very deep black and velvety cosmos, seeing – rather, knowing for sure – that there was a purposefulness of flow, of energy, of time, of space in the cosmos – that it was beyond man's rational ability to understand, that suddenly there was a nonrational way of understanding that had been beyond my previous experience.

There seems to be more to the universe than random, chaotic, purposeless movement of a collection of molecular particles.

On the return trip home, gazing through 240,000 miles of space toward the stars and the planet from which I had come, I suddenly experienced the universe as intelligent, loving, harmonious.

Edgar Mitchell
USA

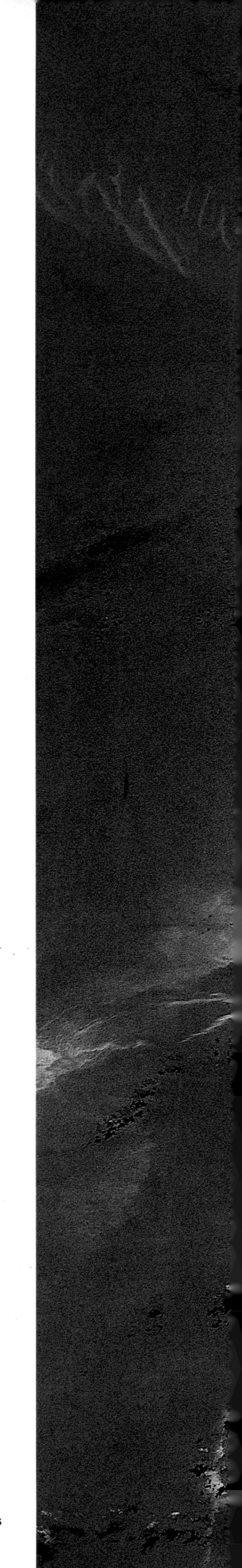

138   Andros Island, Bahamas

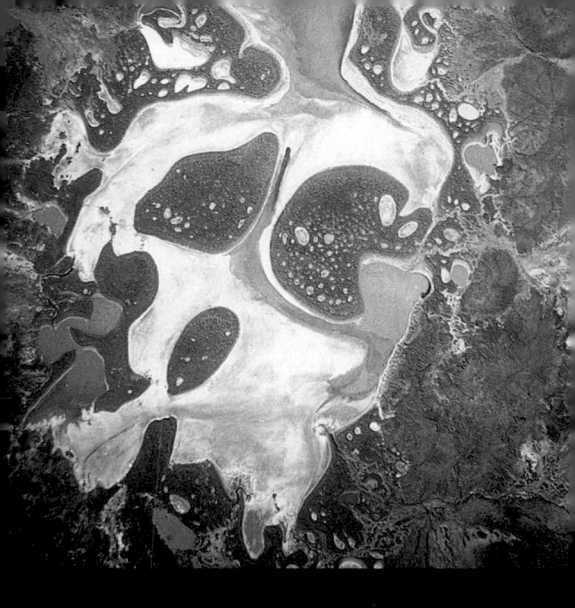

20 Lake Carnegie in western Australia

Мы смотрим в небо, и оно кажется бесконечным. Мы дышим, незамечая этого, как все естественное. Мы, не задумываясь, повторяем: «безбрежный воздушный океан». И вот ты садишься в космический корабль, отрываешься от Земли, и всего за какие-то десять минут проносишься сквозь слой воздуха, за которым – ничего! пустота, холод, мрак. «Безбрежный» голубой океан неба, позволяющий нам дышать, защищающий от бездны и гибели, оказался тонюсенькой пленочкой. Как страшно повредить даже в самой малости эту тонкую оболочку – охранительницу жизни!

Владимир Шаталов
СССР

When we look into the sky it seems to us to be endless. We breathe without thinking about it, as is natural. We think without consideration about the boundless ocean of air, and then you sit aboard a spacecraft, you tear away from Earth, and within ten minutes you have been carried straight through the layer of air, and beyond there is nothing! Beyond the air there is only emptiness, coldness, darkness. The "boundless" blue sky, the ocean which gives us breath and protects us from the endless black and death, is but an infinitesimally thin film. How dangerous it is to threaten even the smallest part of this gossamer covering, this conserver of life

Vladimir Shatalov
USSR

Bereits vor meinem Flug wußte
ich, daß unser Planet klein und
verwundbar ist. Doch erst als ich ihn
in seiner unsagbaren Schönheit und
Zartheit aus dem Weltraum sah,
wurde mir klar, daß der Menschheit
wichtigste Aufgabe ist, ihn fur
zukünftige Generationen zu hüten
und zu bewahren.

Sigmund Jähn
Deutsche Demokratische Republik

Before I flew I was already aware of
how small and vulnerable our planet
is; but only when I saw it from space,
in all its ineffable beauty and fragility,
did I realize that humankind's most
urgent task is to cherish and preserve
it for future generations.

Sigmund Jähn
German Democratic Republic

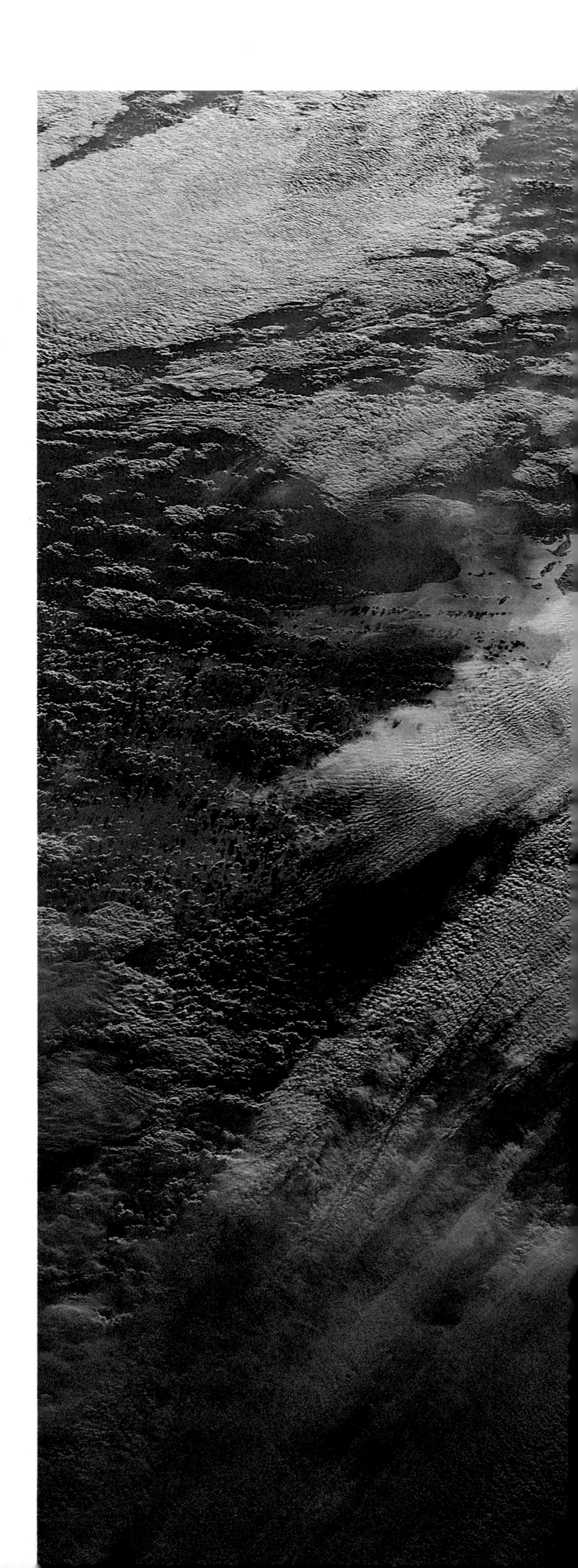

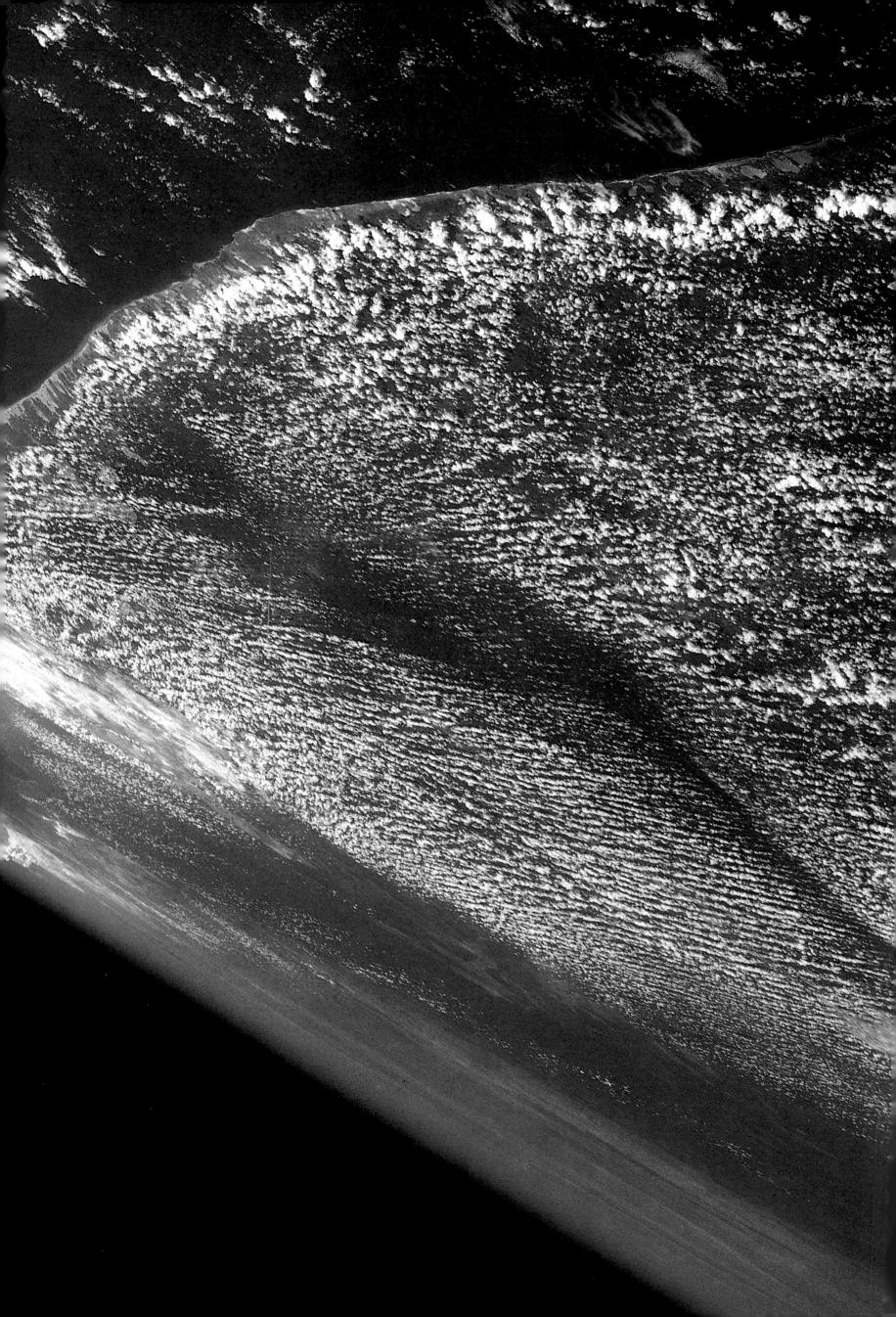

And then you look back on the time
you were outside on that EVA and
on those few moments that you
could take, because a camera
malfunctioned, to think about what
was happening. And you recall staring
out there at the spectacle that went
before your eyes, because now you're
no longer inside something with a
window looking out at a picture. Now
you're out there and there are no
frames, there are no limits, there are
no boundaries. You're really out there,
going 17,000 miles an hour, ripping
though space, a vacuum. And there's
not a sound. There's a silence the
depth of which you've never
experienced before, and that silence
contrasts so markedly with the
scenery you're seeing and with the
speed with which you know you're
moving.

And you think about what you're
experiencing and why. Do you deserve
this, this fantastic experience? Have
you earned this in some way? Are you
separated out to be touched by God, to
have some special experience that
others cannot have? And you know
the answer to that is no. There's
nothing you've done to deserve this, to
earn this; it's not a special thing for
you. You know very well at that
moment, and it comes through to you

so powerfully, that you're the sensing
element for man. You look down and
see the surface of that globe that
you've lived on all this time, and you
know all those people down there, and
they are like you, they are you, and
somehow you represent them. You are
up here as the sensing element, that
point out on the end, and that's a
humbling feeling. It's a feeling that
says you have a responsibility. It's not
for yourself. The eye that doesn't see
doesn't do justice to the body. That's
why it's there; that's why you are out
there. And somehow you recognize
that you're a piece of this total life.
And you're out there on that forefront
and you have to bring it back
somehow. And that becomes a rather
special responsibility, and it tells you
something about your relationship
with this thing we call life. So that's a
change. That's something new. And
when you come back there's a
difference in that world now. There's a
difference in that relationship
between you and that planet and you
and all those other forms of life on that
planet, because you've had that kind
of experience. It's a difference and it's
so precious.

Russell Schweickart
USA

145　The Earth from 10,000 miles

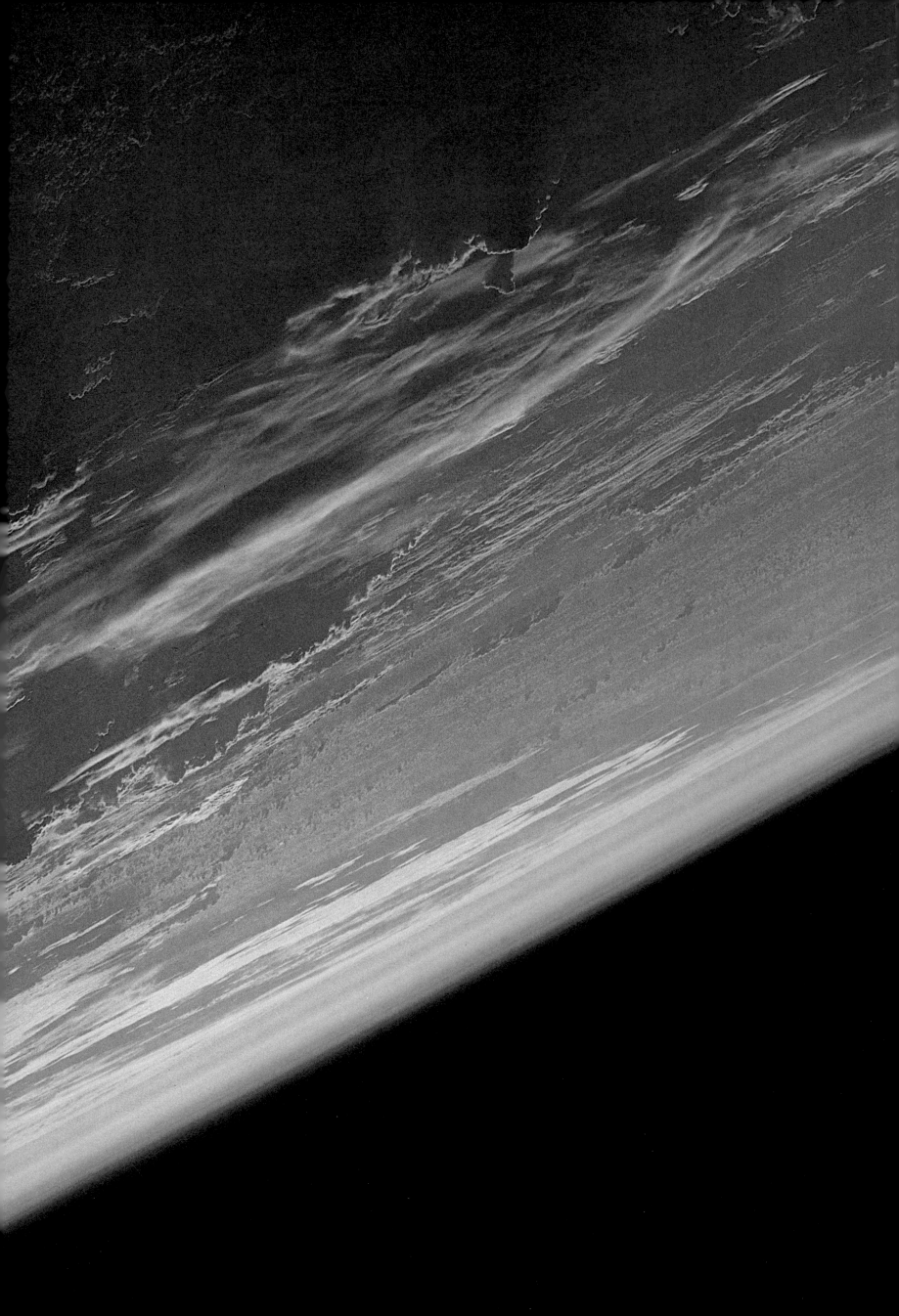

А завтра? Поселения на Луне, путешествия к Марсу, научные станции на астероидах, связь с другими цивилизациями. Не будем огорчаться, что не мы с вами станем участниками дальних межпланетных экспедиций. Не будем завидовать людям будущего. Им, конечно, здорово повезет, для них станет привычным то, о чем мы можем только мечтать. Но и нам тоже выпало большое счастье. Счастье первых шагов в космосе. И пусть потомки завидуют этому нашему счастью.

Юрий Гагарин
СССР

And tomorrow? Settlements on the Moon, voyages to Mars, scientific stations on the asteroids, contact with other civilizations . . . Let us not grieve that we shall not participate in distant planetary expeditions. We shall not envy the people of the future. Of course they are lucky and things about which we can only dream will be ordinary for them. But great happiness has come our way, too, the happiness of the first steps in space. Let those who follow us envy this our happiness.

Yuri Gagarin
USSR

Why Space Photography?

Richard W. Underwood

The first design for the Mercury spacecraft had no windows. They would put a human being inside a container no bigger than an oil drum and shoot him into space. Critics said he had to have a window or he would go crazy. The other side said that a window would be very detrimental to the integrity of the spacecraft from a structural point of view. The window folks won.

Here was the first chance to put a camera on a manned spacecraft. The opponents said that cameras are big and the spacecraft had no room. They weigh a lot, and the engineers were worried about fractions of an ounce. Cameras contain glass and that could be very bad news to astronauts. All the different components would "outgas" and make them very ill, or even kill them. "Cameras are out."

The astronauts saved us. They wanted to have a photo record of the journey and not just a memory recall. The astronauts, thank God, prevailed. Cameras went into space. Astronauts would take tens of thousands of photos of their home planet. I have seen them all.

Before space photography, a traveler walking or driving across the Earth or even in an airplane was like a fly walking across the Mona Lisa at the Louvre in Paris. It would be difficult to appreciate the beauty and the genius of da Vinci that way. You have to back off ten feet to appreciate the masterpiece. The same is true in space. You have to back off a hundred or more miles to see what a masterpiece our home planet is. Thus, we would ask astronauts to take plenty of photographs.

At first, we looked at clouds and the geology of the synoptic view permitted from space. Then we quickly realized that we could see nearly every imaginable geoscientific phenomenon. Thus, a formal photographic plan would develop for each mission and its unique features. The launch time, the orbit inclination, the season, the altitude, and so forth, would all dictate where and when and why we would take the photos.

Geologists could use the photos to see the details of their science, in large areas and over a short period of time, and to learn about the Earth and its dynamics.

The meteorologist could use them to see the development of weather and its movement. Imagine having the ability to track a hurricane from its start to its end.

The oceanographer could see the state of the ocean surface over vast areas, the ocean currents from a new vantage point, the sources and movement of nutrients to feed fish that feed people.

The environmentalist could see the sources and distribution of air and water pollutants and the destruction caused by them. From the smokestack and pipe to acid rain, we can see it all in space photography. Unfortunately, it is a far more air-polluted Earth today than it was in the past, and so the photographs taken by the space shuttle astronauts are not as clear as those taken by Gemini astronauts over twenty years ago.

We knew that, with infrared films, agricultural scientists could use the photos to tell you what grows on the planet and where; how healthy it is; if ill, whether it is a bug or a disease; the date of maturity; the yield rate per area; and even when to irrigate it. They could detect overgrazing, imprudent land use practices, and even illegal activities.

Others who could use it would be the urban planner, the water resource planner, the forester, and many, many more. One can see from some of the photos how, in Africa, over the years, the great Sahara is expanding ten, twenty, even fifty miles a year in some areas, and that the expansion is a threat to the entire population of Africa. But none of the "experts" wanted to look at those photos.

We would even ask astronauts to photograph views that "just look pretty or different or something you just don't understand." Many are contained herewith. Our home planet, from space, can be very beautiful. Occasionally it can be quite ugly as well.

You may wonder why the photos do not represent a comprehensive coverage of the Earth. The Earth is very large, about 197 million square miles. If one space photo covered 10,000 square miles of the Earth's surface, it would take 19,700 of them to cover the Earth with no overlap. Most spacecraft take off nearly straight east from their launch sites, to take full advantage of the Earth's rotation for a boost. That means that US spacecraft, for instance, can only go about 28 degrees north and 28 degrees south, thus covering but a narrow band of the Earth's surface. A number of US missions have been launched with a 50-degree inclination. That way, you see as far north as the southern border of Canada and as far south as Patagonia in Argentina. Almost all Soviet spacecraft are launched at 51.6 degrees. A few spacecraft can go to about a 58-degree inclination – as far north as parts of Alaska or southern Norway and as far south as Cape Horn. To cover the entire Earth, one would have to launch straight south into a polar orbit. When that becomes possible, we will be able to see photos of the Arctic and Antarctic areas; and, uniquely, it will always be the same Earth time (am or pm) beneath the spacecraft. The best is yet to come.

Descriptions
of Photographs

Richard W. Underwood

The photo descriptions that follow are keyed to the number preceding each caption in the text. At the end of these descriptions is the name of the spacecraft and the name and dates of the mission. Except for images from the Apollo missions to the Moon, the photographs were taken in Earth orbit from altitudes ranging from approximately 100 to 400 miles. Primarily, medium-format (2¼″ x 2¼″) Hasselblad cameras were used, with a variety of wide-angle and telephoto lenses. All photographs, except where otherwise noted, are provided courtesy of the National Aeronautics and Space Administration (NASA).

Title page   The full Earth seen from *Apollo 17*, the last journey to the moon, December 1972, from a distance of 23,000 miles. Africa is visible in its entirety. Also visible are most of Antarctica, the Arabian Peninsula, Iran, and a portion of India, as well as the western half of the Indian Ocean and portions of the Atlantic. The cloud formations show movement out of the Antarctic region northward. The band of clouds around the Earth is the Intertropical Convergence Zone, covering areas of rain forests. The sky is clear over the Sahara Desert to the north and the Kalahari Desert to the south. *Apollo Saturn,* AS17, 7 December 1972 – 19 December 1972.

Epigraph   A very thin, crescent Earth, seen from the *Apollo 12* command module as it returned from the moon, November 1969. The brightness in the lower left-hand corner of the photograph is a lens flare caused by sunlight reflecting on the window and the lens. *Apollo Saturn,* AS12, 14 November 1969 – 24 November 1969.

1   Late afternoon, sunset, and night over Africa. The line of light/dark, caused either by the sunrise or the sunset, is called the terminator. The edge of the Earth, or the horizon, is called the limb. In this photograph, limb and terminator blend near the top. *Apollo Saturn,* AS11, 16 July 1969 – 24 July 1969.

2   Three cosmonauts approaching their launch pad at the Kosmodrome Baikonur in the Kazakhstan Republic of the USSR. Photo by A. Makletsov/APN, Novosti Press.

3   Kosmodrome Baikonur, 6 June 1985. The boost rocket, with the spacecraft *Soyuz T-13*, is on the launch pad. Photo by A. Pushkaryov and V. Kuzmin/Fotokhronika TASS.

4   The space shuttle *Atlantis* lifting off the launch pad at Cape Canaveral, Florida. This was NASA's fourth orbiter and carried a five-man crew.

5   Looking due south over portions of South Africa, Namibia, and Botswana, with a thin cloud deck covering north-central South Africa. Part of the Kalahari Desert is visible in the foreground. The desert is a classic red to orange-red, with small depressions and salt encrustations which look like lakes. In the background are Cape Town, the Cape of Good Hope, and the southernmost point of Africa, Cape Agulhas. *Challenger 6*, 41G, 5 October 1984 – 13 October 1984.

6   Just before sunrise over the Pacific Ocean, about 2,100 miles east of Tokyo, Japan. The bands of color show the various layers of aerosols which surround the Earth. The brilliant red layer is the atmosphere; the overlap between the red and blue layers is the stratosphere; the blue layer is the ionosphere. With increased altitude, the electrons and ions are reduced in number, leaving nothing but the blackness of space. *Challenger 6*, 41G, 5 October 1984 – 13 October 1984.

7   The south tip of the island of Madagascar, with sunlight reflecting on the surface of the Mozambique Channel. Mozambique Channel, Madagascar. *Challenger 6*, 41G, 5 October 1984 – 13 October 1984.

8   Seen in long shadows cast by a low Arctic sun, the snow-topped peaks of the Aniakchak Volcano in Alaska's Aleutian Range. *Challenger 6*, 41G, 5 October 1984 – 13 October 1984.

9   Klyuchevskaya Volcano. At 24,500 feet, the volcano is the highest of the great volcanos of the Kamchatka Peninsula in Siberia, USSR. The Kamchatka River flows halfway around the volcano's north side. The town of Klyuchi is at the top center edge of the photograph. *Challenger 9*, 61A, 30 October 1985 – 6 November 1985.

10   A long trail of clouds is being drawn into the vortex of tropical storm "Xina," north of the Hawaiian Islands. A vortex in the Northern Hemisphere always moves counter-clockwise; in the Southern Hemisphere the airflow is clockwise. This is called the Coriolis effect and is caused by the rotation of the Earth on its polar axis. *Challenger 9*, 61A, 30 October 1985 – 6 November 1985.

11   Typhoon "Odessa," about 1,200 miles northeast of Hawaii. The storm vortex, seen in its entirety, moves counter-clockwise with a very well-defined eye. *Discovery 6*, 51I, 27 August 1985 – 3 September 1985.

12   Cape Cod in Massachusetts, seen against a very blue Atlantic Ocean. The cape is a deposit of earth and stone, called a terminal moraine, left by the great Pleistocene glaciers of about 20,000 years ago. *Challenger 6*, 41G, 5 October 1984 – 13 October 1984.

13   The spacecraft *Challenger*, photographed by a remote radio operated Hasselblad, 70mm camera attached to the West German-built SPAS unmanned spacecraft. *Challenger 2*, STS7, 18 June 1983 – 24 June 1983.

14   Sunset over the coastal ranges of Yukon and British Columbia in Canada and southeastern Alaska, near Mount Saint Elias. The view is to the northwest into an Arctic night. *Challenger 6*, 41G, 5 October 1984 – 13 October 1984.

15   Long Island in the Bahamas (running from lower right to upper left), Little Exuma (at the right edge below center), with assorted small reefs and a portion of Great Exuma Island. The dark blue is the very deep waters of Exuma Sound and Crooked Island Passage. One can see to depths of several hundred feet in these pristine waters. *Challenger 4*, 41B, 3 February 1984 – 11 February 1984.

16   A vortex over the mid-Indian Ocean. *Challenger 6*, 41G, 5 October 1984 – 13 October 1984.

17   The Bahama Islands southeast of Florida, with contrasting shallow water, exposed and subsurface reefs, islands, deep sounds, and clouds. The northernmost larger islands of Grand Bahama and Great Abaco are near the center of the photograph with a portion of Andros Island at the upper right, and the Bimini Islands at right of center. New Providence Island, on which can be seen Nassau. the capital, is the small island near the upper right corner. *Challenger 7*, 51B, 29 April 1985 – 6 May 1985.

18   The Altiplano and the great Andean volcanos northeast of Salar de Atacama, where Argentina, Bolivia, and Chile meet. The small green lake is appropriately called Laguna Verde. *Columbia 7*, 61C, 12 January 1986 – 18 January 1986.

19   Fjord-like lakes carved by glaciers on New Zealand's South Island. At the center of the photograph is the azure Lake Pukaki, with Lake Tekapo near the upper edge of the photograph and Lake Ohau near the bottom. *Challenger 9*, 61A, 30 October 1985 – 6 November 1985.

20   The coast of California. The coastline is seen from Tomales Bay southward, past Point Reyes and Bolinas, the Golden Gate and Half Moon Bay, to just north of Santa Cruz, as seen from shuttle mission 61A, the second mission for the European Space Agency *Spacelab*. Along the right edge of the photograph is a portion of the San Joaquín and Sacramento valleys. The unique color patches at the lower end of San Francisco Bay (near Fremont) and at the upper end of San Pablo Bay (near Vallejo) are salt evaporators. The sharp gash in the earth which crosses the photo is the very active San Andreas Fault. *Challenger 9*, 61A, 30 October 1985 – 6 November 1985.

21   Southeastern Algeria, about 1,000 miles south-southeast of Algiers, between Bordj-onar-Driss and the city of Tamanrasset. Near the center of the photograph is an area of star-shaped dunes. This classic shape is caused by winds blowing in all directions and forcing the sand upward. *Columbia*, STS2, 12 November 1981 – 14 November 1981.

22   Captain Bruce McCandless using the Manned Maneuvering Unit (MMU) for the first unattached space walk on 7 February 1984. A 35mm Nikon camera is mounted to the right of his head. *Challenger 4*, 41B, 3 February 1984 – 11 February 1984.

23   Russell Schweickart peeks around the corner of the lunar module on *Apollo 9*. *Apollo Saturn*, AS9, 3 March 1969 – 13 March 1969.

24   Edward H. White II during his 3 June 1965 EVA (extra-vehicular activity). He holds the maneuvering unit in his right hand. *Gemini 4*, 3 June 1965 – 7 June 1965.

25   Cosmonaut Leonid Kizim works in open space during *Salyut 7* mission. Photo by V. Solovyov/Fotokhronika TASS. *Soyuz T-10*, 8 February 1984 – 21 October 1984.

26   Svetlana Savitskaya performing a welding experiment during the first space walk by a woman. Photo by V. Dzhanibekov/ Fotokhronika TASS. *Soyuz T-12*, 17 July 1984 – 29 July 1984.

27   Dawn gives rise to spectacular colors. (See Photo 6.) *Challenger 6*, 41G, 5 October 1984 – 13 October 1984.

28   Photograph of dawn taken from manned orbital complex *Soyuz 26*, December 1976. Photo taken by Cosmonauts G. Grechko and Yu. Romanenko on board the space station. Fotokhronika TASS. *Soyuz 26*, 10 December 1977 – 16 March 1978.

29 Sunset, lit by Earth's atmosphere. Photo taken by Cosmonauts V. Kovalyonok and A. Ivanchenkov on board the orbital space complex *Salyut 6 - Soyuz 29 - Soyuz 31*, September 1978. Fotokhronika TASS. 15 June 1978 – 2 November 1978.

30 The Earth's limb and terminator over Southeast Asia seen from nearly 1,000 miles away. Glints from the sun catch the high clouds to form the golden crown of atmospheric layering. *Challenger 7*, 51B, 29 April 1985 – 6 May 1985.

31 An aurora borealis over the Southern Hemisphere, called an Aurora Australiasis. To the right side of the red-capped main aurora are two thin layers in the very thin upper atmosphere. The higher layer is a diffuse aurora, only half a mile thick, which was not accounted for in aurora theory until this photograph proved its existence and raised many, many new questions about the cause and mechanics of these phenomena. The lower layer, also seen on the left, is the air glow layer through which you can see stars shining. The air glow layer lies at about 55 miles above the Earth, and is activated by ultraviolet light from the sun. As that part of the atmosphere rotates into the Earth's shadow it remains luminescent for several hours. *Challenger 7*, 51B, 29 April 1985 – 6 May 1985.

32 – 34 A moonrise series taken from *Skylab 3*. These pictures proved the effect of atmospheric distortion on a perfect sphere. Viewed through the denser air near the Earth's surface, the moon appears flattened. As it rises above the atmosphere it regains its spherical shape. *Skylab 3*, SL3, 28 July 1973 – 25 September 1973.

35 The eastern "far side" of the moon as seen from *Apollo 11*. This rugged terrain, viewed to the southwest, is typical of the "far side" which is exposed to meteor bombardment. The side facing Earth is more protected, lying as it does in our gravitational "lee." International Astronomical Union crater 308 is clearly visible, located at 179 degrees E, 5 degrees S. The crater is about 50 miles in diameter. *Apollo Saturn*, AS11, 16 July 1969 – 24 July 1969.

36 The island of Hawaii, which originally was two islands with two distinct types of volcanos. The continual lava eruptions of Mauna Loa eventually created a larger, single island by joining with Mauna Kea, a violent bombastic volcano that ejected rock and dust. *Challenger 9*, 61A, 30 October 1985 – 6 November 1985.

37 The coastline of southern Africa seen from *Apollo 11* shortly before splashdown in the Pacific Ocean. *Apollo Saturn*, AS11, 16 July 1969 – 24 July 1969.

38 A nearly full disk of the Earth. Visible are the west coast of North America, from Alaska southward to Panama, a bit of northwest South America, Baja California, the Great Basin, and the snow and ice covering northern Canada, Hudson Bay, and Greenland. *Apollo Saturn*, AS10, 18 May 1969 – 26 May 1969.

39 A full moon seen from *Apollo 11* as it moved away from the lunar surface early in its trans-Earth trajectory. The right third of the surface seen in this photograph is "far side" and cannot be seen from the Earth. *Apollo Saturn*, AS11, 16 July 1969 – 24 July 1969.

40 The far side of the moon, at a point exactly opposite the center of the lunar plain which we see from Earth. A small portion of International Astronomical Union crater 308 is seen at upper right. *Apollo Saturn*, AS11, 16 July 1969 – 24 July 1969.

41 The moon as viewed from *Apollo 17*. In the center of the photograph is the Thomson crater, about 90 miles in diameter, located about 32 degrees S, 166 degrees E. The shadow at the left is the southwest wall of crater Van der Graf. The lower right shadow is a portion of the crater Zelinsky. *Apollo Saturn*, AS17, 7 December 1972 – 19 December 1972.

42 – 45 In this series, which was taken from *Apollo 11* while in lunar orbit, 20 July 1969, a cloud-covered Earth appears to rise above the surface of the moon. The entire continent of Australia is visible, as is the western Pacific Ocean, eastern Asia, and the Arctic. *Apollo Saturn*, AS11, 16 July 1969 – 24 July 1969.

46 The *Apollo 17* "Moonwalkers" Eugene Cernan and Jack Schmitt arrive at Station 6 and photograph Boulder 2 in the valley of Taurus-Littrow. Station 6 is about 2 miles north of the landing site and at the base of the North Massif. Boulder 2, about 6 feet in height, is composed mainly of a vesicular blue-gray and green-gray breccia with small white clasts. The dust at the bottom surface is a shedding effect caused by bombardment by small space particles and also by the quick and severe temperature changes (from 300°F to −450°F) from sunlight to shadow. *Apollo Saturn*, AS17, 7 December 1972 – 19 December 1972.

47 An *Apollo 17* view at Station 5 in the valley of Taurus-Littrow, looking toward the East Massif. The rock in the foreground is about 3 feet tall. *Apollo Saturn*, AS17, 7 December 1972 – 19 December 1972.

48 Astronaut Jack Schmitt about to walk behind a large boulder nicknamed "Split Rock." The boulder is probably excecta material thrown up from some other large lunar impact. *Apollo Saturn*, AS17, 7 December 1972 – 19 December 1972.

49 The Earth seen from Taurus-Littrow landing site of *Apollo 17*. The large moonrock in the foreground is about 6 feet tall. *Apollo Saturn*, AS17, 7 December 1972 – 19 December 1972.

50 A very slim crescent Earth seen by the *Apollo 11* crew on 23 July 1969, the day before splashdown. *Apollo Saturn*, AS11, 16 July 1969 – 24 July 1969.

51 The view of Earth from *Apollo 15* en route to the moon. At the upper right of the photograph are a portion of Spain and western Africa. The north Atlantic is at upper center; the south Atlantic is at lower right. North America is at upper left and South America is at lower center. The desert lands of Peru, Bolivia, Argentina, and Chile stand out in buff. *Apollo Saturn*, AS15, 26 July 1971 – 7 August 1971.

52 Africa, the Near East, the Mediterranean Sea, Europe, and the western half of the USSR, seen from *Apollo 11* during "translunar coast," 16,000 miles from Earth. *Apollo Saturn*, AS11, 16 July 1969 – 24 July 1969.

53 Planet Earth seen from *Apollo 12* returning home. The sun is eclipsed at the edge of the Earth's limb. The photograph is one of a series taken by a 16mm data-recording camera. *Apollo Saturn*, AS12, 14 November 1969 – 24 November 1969.

54 A crescent Earth rises over the moon, as seen by astronauts in lunar orbit. The exposure was set for the lunar surface and thus the Earth appears overexposed. *Apollo Saturn*, AS15, 26 July 1971 – 7 August 1971.

55 The moon viewed to the southeast from *Apollo 17*, the last moon landing. Mare Fecunditatis is in the upper left. To the upper right is Mare Serenitatis. *Apollo Saturn*, AS17, 7 December 1972 – 19 December 1972.

56 A very slim crescent Earth as seen on the return trip from the moon. Clouds often take on a golden cast near the terminator at sunrise or sunset. This is the cause of the bright golden spot at the center of the crescent. *Apollo Saturn*, AS12, 14 November 1969 – 24 November 1969.

57 Africa, Arabia, parts of Iran, India, and Pakistan, along with all of Madagascar Island and the western half of the Indian Ocean. The camera, modified for lunar flight, had no viewfinder, requiring a "shoot from the hip" technique. *Apollo Saturn*, AS17, 7 December 1972 – 19 December 1972.

58 The Kyzyl Kum Desert of the Kazakh and Uzbek Soviet Socialist Republics. The Aralskoe More, also known as the Aral Sea, is to the south-southeast. This saltwater lake is fed mainly from waters of the Amu Darya River, mostly with snowmelt water from the distant Tyan-Shan Mountains near the Afghanistan and China border. *Challenger 6*, 41G, 5 October 1984 – 13 October 1984.

59 "Flatirons," cumulonimbus clouds that have flattened out at a high altitude, the result of rapidly rising moist air. At a given altitude, depending on temperature, wind, and humidity, the cloud mass can no longer rise and the wind aloft shears the cloud. Central Nigeria, an area in which tropical rain forest gives way to dryer savannah lands, lies beneath a layer of heavy haze and smoke. *Challenger 6*, 41G, 5 October 1984 – 13 October 1984.

60 Low clouds over the Atlantic coast of South Africa and Namibia near Port Nolloth. Despite the fact that some places in this area get less than an inch of rain a decade, clouds and fog often cover the coastline as the cold, north-flowing Benguela Current comes in contact with the hot deserts. The plants and wildlife adapt to these unique conditions of clouds and fog without rain by removing moisture for survival directly from the atmosphere. *Atlantis 2*, 61B, 26 November 1985 – 3 December 1985.

61 Richat, a hole 20 miles across and 2,000 feet deep in Mauritania, West Africa. First identified by World War II aircraft and believed to be a meteorite impact area, it is in fact a wind erosion residual. Sahara winds traveling at very high speeds move sand around the structure, causing abrasive erosion and continuing enlargement. *Discovery 2*, 51A, 8 November 1984 – 16 November 1984.

62 The Gulf of Carpentaria between Australia's states of Queensland and Northern Territory. The gulf has very shallow water over a submerged coastal plain, making the sea bottom visible under certain conditions, particularly if the surface is light sand or coral. The coastline of western Queensland is visible, as are the Wellesley Islands. *Atlantis 2*, 61B, 26 November 1985 – 3 December 1985.

63 Eighty Mile Beach, where the Great Sandy Desert reaches the Indian Ocean coast of western Australia. The northwest coastal highway crosses the area past Sandfire Flat Roadhouse. The nearest towns are Broome, nearly 150 miles northeast, and Port Headland, nearly 150 miles to the west. Salt flats, sand ridges, and a few rocky outcrops mark this very remote area. *Discovery 1*, 41D, 30 August 1984 – 5 September 1984

64 A large storm in the North Pacific Ocean south of Alaska. The predominantly white area is the Kamchatka Peninsula of Siberia. Japan appears near the horizon. *Apollo Saturn*, AS13, 11 April 1970 – 17 April 1970.

65 The Florida Strait between Florida and the Bahamas, with Gulf Stream surface currents shown in the sparkling sunlight. The edge of the Great Bahama Bank and the sea bottom along the west side of Andros Island are clearly defined. *Gemini 4*, 3 June 1965 – 7 June 1965.

66 A plankton bloom in an upwelling current reveals a distinct Southern Hemisphere clockwise vortex in coastal waters off New Zealand's South Island, northeast of Christchurch. The "bloom" moves northeastward in currents that impinge on Banks Peninsula, which causes the downstream vortex seen here off Pegasus Bay. *Challenger 9*, 61A, 30 October 1985 – 6 November 1985.

67 The Indian Ocean near Madagascar. The glint of the sun through atmospheric pollution (smoke) causes an intense gold color rarely seen by astronauts. *Discovery 5*, 51G, 17 June 1985 – 24 June 1985.

68 A portion of the Tongue of the Ocean and the Great Bahama Bank. The light blue color is the shallow waters (10 to 200 feet deep) of the bank; the medium blue is water at a depth of up to 400 feet. The abrupt change from the lighter blues to the dark blue of the Tongue of the Ocean indicates an underwater cliff dropping off to a depth of about one mile. The color differentials at the edge of the cliff are submerged canyons cut into the coral. These are kept open by the great hurricanes which permit water to transfer from the shallow to the deeper areas. *Columbia 5*, STS5, 11 November 1982 – 16 November 1982.

69 The Blue Nile at Wad Medani in Sudan appears at the upper left of the photograph. The lesser meandering stream to the right is Nahar Ar Rahad. The only cash export crop of the region is cotton, and the only reliable source of water is the Nile. Solar reflection from the irrigation project at the lower center reveals the complexities of the canal delivery system.

70 The eastern half of the Indonesian island of Java, with the islands of Bali, Lombok, Madura, and Sumbawa farther to the east. Great volcanic peaks rise to form the island chain. Mahameru Volcano near the center of the photograph rises to over 19,400 feet. *Challenger 7*, 51B, 29 April 1985 – 6 May 1985.

71 In southern Brazil, a large kidney-shaped lagoon called Lagoa Dos Patos. It was created by a 500-mile-long barrier beach, in turn formed by north-moving coastal currents. Suspended sediment in the shallow, mud-filled lagoon is mixed with raw sewage from Porto Alegre, a city of two million people, just off right center at the upper edge of the photograph. Blue water is visible in one isolated portion of the lagoon, probably due to recent runoff. *Skylab 3*, SL3, 28 July 1973 – 25 September 1973.

72 Center-pivot irrigation wells in fields in the Brazilian state of Mato Grosso Do Sul, south of Campo Grande. The water is broadcast in circular patterns. *Challenger 6*, 41G, 5 October 1984 – 13 October 1984.

73 Sun glints on the small and very numerous glacial lakes of the Canadian Shield of Quebec. This is an area of very ancient pre-Cambrian rock, which uplifted when the weight of the Pleistocene glaciers was removed. The double crescent lake is Lake Mistassini, some 400 miles north of Montreal. *Challenger 9*, 61A, 30 October 1985 – 5 November 1985.

74 Madagascar's Betsiboka River flowing into Majunga Bay, which is quickly being silted in due to imprudent agriculture and forest practices. *Discovery 2*, 51A, 8 November 1984 – 16 November 1984.

75 The head of the Persian (or Arabian) Gulf, Iran, and the great ridges of the Zagros Mountains in the foreground. Iraq and Kuwait are in the background, a large dust storm partially obscuring some areas. Smoke from battle-related fires – including a burning ship in the gulf itself – is visible at several locations in this area of the so-called Gulf War. *Challenger 6*, 41G, 5 October 1984 – 13 October 1984.

76 Dasht-i-Kavir, in Iran, southeast of Tehran, as seen from the first mission of the space shuttle *Columbia*. Dasht-i-Kavir (meaning a salt and gravel encrusted pediment) is a windblown desert floor of salt flats and salt lake. *Columbia*, STS1, 12 April 1981 – 14 April 1981.

77 A southwestern view of the Greater Himalayas bordering the Karakoram Range. India is to the left, Pakistan to the right, and China is in the foreground. The valley of the Indus River is in the right background and the fabled Vale of Kashmir is near the right edge of the photograph. The great peaks of the Karakoram Range are near the lower edge. *Challenger 6*, 41G, 5 October 1984 – 13 October 1984.

78 The sun reflects through an unusual pattern of linear clouds over a swampy area of the state of Corrientes, Argentina, near the Rio Uruguay. *Discovery 6*, 51I, 27 August 1985 – 3 September 1985.

79 Long parallel shadows at sunset across the flat plain of the Chaco region of Paraguay. Bright highlights fill the higher clouds, which catch the waning sunlight. *Discovery 2*, 51A, 8 November 1984 – 16 November 1984.

80 The Strait of Hormuz, separating Iran in the background from the more colorful Arabian Peninsula. The Muscat and Trucial coasts are to either side of the strait. The elongated ridges of southern Iran's Zagros Mountains are in the background. *Challenger 9*, 61A, 30 October 1985 – 6 November 1985.

81 The eastern Mediterranean Sea, with the movement of surface currents around the various islands in the Aegean Sea clearly visible in the reflected sunlight. Currents like these can be studied on such a large scale only from orbiting spacecraft. The long island to the left and center of the photograph is Crete. The Peloponnesus Peninsula of Greece and the City of Athens are to the right. The African coast of Libya and Egypt and the northern Sahara are toward the horizon, with Turkey in the foreground. Islands from right to left along the Turkish coast include Lesbos, Chios, Samos, Cos, and Rhodes near the lower left. *Challenger 6*, 41G, 5 October 1984 – 13 October 1984.

82 The flat Pampas of Argentina's Atlantic coast, on the lower right, with the Rio de la Plata at the lower center of the photograph. The background clouds hover over the South Atlantic and Patagonia. *Challenger 6*, 5 October 1984 – 13 October 1984.

83 Southwestern Algeria's Erg Chech shows long lines of parallel sand dunes called siefs. The Erg (sand desert) is in a remote area (26.5 degrees N, 1.5 degrees W) of harsh desert, uninhabited and rarely visited. These parallel sand dunes are about 100 miles in length and 5 to 10 miles apart and are found in very few areas of the Earth. Most sand dunes are transverse dunes, or perpendicular to the general direction of the wind. *Challenger 5*, 41C, 6 April 1984 – 13 April 1984.

84 The deeply etched patterns of intermittent streams in the Hadhramaut Plateau of South Yemen. These stream beds, which now rarely carry water, are deeply entrenched, owing to the geologically recent uplift of the plateau. Nonetheless, a very distinct watercourse appears along the top and bottom edges of the photograph. In the photograph's upper portion, available water will flow into the vast emptiness of the Rub al Khali (Empty Quarter). In the lower portion, water will flow to the Gulf of Aden. *Challenger 6*, 41G, 5 October 1984 – 13 October 1984.

85 The Gambia River, from Georgetown in the upper right corner of the photograph downstream to Mansa Konko, where it becomes an estuary about 60 miles above Banjul, the capital of The Gambia. The river is heavily silted from rain originating in the highlands of Guinea. This accounts for its bronze color as it passes from forest to drier savannah en route to the Sahara and across the nation of Senegal in West Africa. The Gambia itself is a long, narrow nation which follows the river for about 200 miles. *Discovery 1*, 41D, 30 August 1984 – 5 September 1984.

86 Lake Van in eastern Turkey, southwest of Mount Ararat. Its water, trapped by mountains, bitter and undrinkable, was an ancient source of soap. The sunlight on the lake reveals the pattern of surface currents. It also illuminates the Buhtan River and its tributaries, which drain away from the Satak Daglari Mountains forming the southwestern watershed of the lake, and is one of the sources of the Tigris River. *Challenger 6*, 41G, 5 October 1984 – 13 October 1984.

87 The snow-covered Belcher Islands in Canada's Hudson Bay off the coast of northern Quebec. Long, low, flat, sandy, and undulating, the islands are the unique result of Hudson Bay currents depositing the very light silt from past continental glaciation. *Challenger 6*, 41G, 5 October 1984 – 13 October 1984.

88 Rodrigues Island, a volcanic outcrop in the Indian Ocean about 1,000 miles east of Madagascar. Discovered by the Portuguese in 1645, it is part of Mauritius, whose main islands are about 300 miles to the west-southwest. Mount Limon, the high central peak, reaches an elevation of 1,300 feet. The coral reef is to the west and south sides of the island, which is itself sinking very slowly into the ocean. As the island sinks, the coral grows to form a vertical wall several thousand feet high. Many volcanic islands in the Southern Hemisphere have sunk below sea level, leaving only coral atolls with vertical sides, some rising over 20,000 feet to the ocean surface. *Discovery 6*, 51I, 27 August 1985 – 3 September 1985.

89 Aswan Dam, high on the Nile River, which formed Lake Nasser in Upper Egypt. Below the dam, the river is in a narrow valley of dark irrigated fields. The escarpment of the Libyan Desert appears in the upper right section of the photograph. The complex volcanic formations of the Eastern Desert are to the left. *Challenger 6*, 41G, 5 October 1984 – 13 October 1984.

90 The *Soyuz T-2* spacecraft orbiting the Earth. The photo was taken by Cosmonauts Leonid Popov and Valeri Ryumin on board the space station *Salyut-6 – Soyuz 35* after separation from the *Soyuz*. Fotokhronika TASS. *Soyuz 35*, 9 April 1980 – 11 October 1980.

91 The Pacific Ocean, several hundred miles southwest of San Diego. The ocean's temperature, its currents, and the upwelling of water at various temperatures cause the two completely different cloud formations. The cellular pattern of stratocumulus clouds in the background covers cool offshore waters. *Gemini 12*, 11 November 1966 – 15 November 1966.

92 Massive, deep canyons and volcanic peaks south of the Peruvian city of Cuzco. The Andes, longest of mountain ranges, forms an unbroken wall between Santa Marta in northern Colombia, where peaks rise 19,000 feet out of the Caribbean Sea, and Tierra del Fuego, over 5,000 miles to the south. *Columbia 7*, 61C, 12 January 1986 – 18 January 1986.

93 A light, thin cirrus cloud layer partially obscures the island of Gran Canaria in the Canary Islands (15.5 degrees W, 28 degrees N). *Challenger 6*, 41G, 5 October 1984 – 13 October 1984.

94 In Namibia, the round dark intrusion known as Brandberg, 20 miles in diameter, rises up abruptly some 8,600 feet. At lower right is another intrusive called Ebongoberge, with a 7,200-foot elevation and a 30-mile diameter. These structures are formed when intense internal pressures force molten rocks to the surface, accompanied by chemical alterations of rocks and ores due to intense heat. Small clouds centered above the Brandberg peaks are typical, and owe their existence to the differences in temperature, humidity, and pressure at higher altitudes. *Atlantis 2*, 61B, 26 November 1985 – 3 December 1985.

95   The complex coastline of Brazil's Paraná state and the port of Paranagua. The long coastal island of Comprida and the Rio Ribeira Do Iguape are outlined in the glint of sun at the top left of the photograph. *Challenger 1*, STS6, 4 April 1983 – 9 April 1983.

96   The mouths of the Ganges in Bangladesh, a massive drainage area forming many channels and islands as it empties into the Bay of Bengal. When great typhoons move ashore, this highly populated area is totally inundated, and often tens of thousands lose their lives. Yet survivors will always return, because the rich delta soil is so highly productive. *Atlantis 2*, 61B, 26 November 1985 – 3 December 1985.

97   The deep blue Gulf of Suez, separating Sinai (bottom) from the Eastern Desert of Egypt (above). In this area "continental drift" is rapid, the two shorelines moving apart about four inches each year. The Great Rift Zone of East Africa extends northward to become the Red Sea and splits into two major faults, one the Gulf of Suez, the other the Gulf of Aqaba and the Dead Sea. *Atlantis 2*, 61B, 26 November 1985 – 3 December 1985.

98   Niger, in West Africa. Man-made drought caused the Sahara Desert to expand into this area. Once savannah lands, it is now primarily sand and dunes and small salt flat lakebeds. Dillia, a wash, occasionally has flowing water, but far less than only a few years ago. *Discovery 4*, 51D, 12 April 1985 – 19 April 1985.

99   Both parallel and transverse dunes, as well as areas of wind-scoured bedrock, in the Namib Desert in Namibia. This very dry land has a unique flora and fauna adapted to the unusually harsh conditions of the desert. *Discovery 1*, 41D, 30 August 1984 – 5 September 1984.

100   A clearly defined vortex over the Indian Ocean, 1,500 miles west of Australia. *Challenger 9*, 61A, 30 October 1985 – 6 November 1985.

101   The eye of tropical storm "Blanca" developing and organizing into a weather system off the west coast of Mexico. *Discovery 5*, 51G, 17 June 1985 – 24 June 1985.

102   The long, linear parallel ridges of the Zagros Mountains of southwestern Iran. Dark, round salt domes intrude from deep beneath the earth to produce oil, much of which has yet to be exploited in this area. *Challenger 6*, 41G, 5 October 1984 – 13 October 1984.

103   The Nile River in Egypt, just below Luxor, near the Valley of the Kings, where the pharaohs were buried. The ancient city of Thebes is along the Nile at the very edge of the photograph. *Challenger 6*, 41G, 5 October 1984 – 13 October 1984.

104   Spring thaw along Canada's Hudson Bay coast of Quebec. The ice is starting to break up, and several months later ships will load wheat from the Hudson Bay ports of Churchill and Moosonee. The long, dark swirls are the low sandy Belcher Islands, a remnant of the continental glacial age. (See Photo 87.) *Challenger 7*, 51B, 29 April 1985 – 6 May 1985.

105   The Bay of Bengal coast of Burma, showing Ramree Island and the Arakan Yoma Mountain Range. This 200-mile stretch of Burma coastline is made up of tidal estuaries, islands and mangrove swamps. Just inland are the mountains, where tremendous rainfall fills the short rivers and estuaries with silt. *Skylab 4*, SL4, 16 November 1973 – 8 February 1974.

106   The Plateau du Djado in the north of Niger, eroded by a wash called the Blaka. This river is creating a new eroded surface in the recently uplifted plain. *Challenger 4*, 41B, 3 February 1984 – 11 February 1984.

107   The Gascoyne River, an intermittent large wash 500 miles north of Perth, Australia, which flows for nearly a thousand miles across the Macadam Plains. The scant water available in this area either is absorbed in the sands and gravel of the watercourse or quicky evaporates. Most water flows beneath the surface, then, giving support to plant root systems and sustaining green vegetation along the watercourses. *Discovery 1*, 41D, 30 August 1984 – 5 September 1985.

108   The province of Manitoba, Canada, at the north end of Lake Winnipeg. This subarctic area was severely glaciated in the not-too-distant geologic past. The slim point of land, Limestone Point, nearly closes off Limestone Bay from the rest of the lake. The town of The Pas is at the bottom left edge of the photograph. NASA large-format camera (LFC) imagery 41G-949 distributed under contract to NASA by Chicago-Area Survey, Inc., 2140 Wolf Road, Des Plaines, IL 60018. *Challenger 6*, 41G, 5 October 1984 – 13 October 1984.

109   Snow cover in the Desolation Canyon area along the Green River of Utah. The Roan Cliffs and East Tavaputs Plateau stand out. The uplift of the Roan and Tavaputs area in the recent geologic past caused the river canyons to dig deeper. *Skylab 4*, SL4, 16 November 1973 – 8 February 1974.

110   Lake Powell of Arizona and Utah, seen in winter from *Skylab 4*. The lake, held back by the Glen Canyon Dam, 710 feet high, was named for John Wesley Powell. In 1869 Powell led the first expedition down the Green-Colorado river system from Green River, Wyoming, through Glen Canyon, and later through the Grand Canyon to the vicinity of Las Vegas, Nevada. The Colorado River appears from lower Canyonlands National Park downstream to the upper Grand Canyon National Park. *Skylab 4*, SL4, 16 November 1973 – 8 February 1974.

111   Cellular cloud structures above the Atlantic Ocean, east of Ascension Island. This is the kind of cloud structure found in stable air over a stable ocean, with the air traveling in slow convective movements. This phenomenon is typical for large areas of ocean between Hawaii and the West Coast of the USA. *Discovery 5*, 51G, 17 June 1985 – 24 June 1985.

112   Poluostrov Shipunskiy on the east coast of the Kamchatka Peninsula juts into the North Pacific Ocean from the east coast of Siberia. The dark blue is typical of ocean water at near-freezing temperatures. *Columbia 6*, STS9, 28 November 1983 – 8 December 1983.

113   The island of Hawaii, with Maui, Kahoolawe, Lanai, and Molokai to the left. The windward clouds have a defined cellular structure, whereas the leeward clouds have been disturbed by their passage over the islands. *Discovery 5*, 51G, 17 June 1985 – 24 June 1985.

114   Northwest of the city of Manaus, Brazil, the Rio Negro meanders through tropical rain forest in the Amazon Basin. The dark, clear waters of the Rio Negro will meet the muddy Amazon 100 miles southwest. The mixing of the two types of water can be seen from space for over 100 miles downstream. *Columbia 7*, 61C, 12 January 1986 – 18 January 1986.

115   The Persian Gulf, looking southwest, with Iraq and Iran and the Shatt-al-Arab River seen at the lower edge of the photograph from al-Basrah past Abadan. Kuwait is to the lower right, with Saudi Arabia along the right edge of the photograph. Iran covers the left side, showing its long gulf coast, as well as the long ridges of the Zagros Mountains. Right of center on the far horizon is the Strait of Hormuz. *Challenger 6*, 41G, 5 October 1984 – 13 October 1984.

116 Reflected sunlight gives the surface of the Pacific Ocean a bronze hue near Arno Atoll (bottom center) in the Ratak, or eastern chain of atolls of the Marshall Islands. The atolls are coral build-ups around submerged mountain range peaks. *Discovery 6*, 51I, 27 August 1985 – 3 September 1985.

117 The Manicouagan Impact, 50 miles in diameter and 300 miles north-northwest of Quebec City. The structure was left by a massive meteorite collision in the far distant past. Lake Manicouagan, created by the Daniel Johnson Dam, is covered by ice. *Challenger 7*, 51B, 29 April 1985 – 6 May 1985.

118 The west coast of the Indonesian portion of the island of Borneo. In the lower center of the photograph is the city of Pontianak; silt enters the sea from the Kapuas River. Small puffs of cumulus clouds tend to form over the jungle rain forests, but not over open water. Glints from the sun show the complexity of the coastal mangroves. *Atlantis 2*, 61B, 26 November 1985 – 3 December 1985.

119 The Indonesian portion of the island of New Guinea, an area now known as Irian Jaya. The body of water is Teluk Berau, or MacCluer Gulf. This is an area of immense tropical rain forest, very heavy, nearly continuous rainfall, and high, steep mountains adjacent to embayments. This produces the muddy water and mangrove swamps evident in the photograph. *Challenger 7*, 51B, 29 April 1985 – 6 May 1985.

120–122 A sequence of three photos taken as the shuttle *Atlantis* moved along the northeast coast of Africa.
In the first photograph we see the Horn of Africa and the junction between the Gulf of Aden and the Indian Ocean. Tilting of the Earth's crust formed the mile-high escarpment that separates the Arabian Peninsula from Africa, which is now filled by the Gulf of Aden (top of photograph). The horn is also called Raas, or Cape, Caseyr. Somalia occupies most of the photograph, with a bit of Ethiopia to the left. Along the east coast of Somalia, strong currents in the Indian Ocean have connected the island of Raas Xaafuun to the mainland.
The second photograph was taken farther down the coast. We still see a portion of the Gulf of Aden to the upper left, with the Indian Ocean to the right, but more of Ethiopia.
The Ogaden Desert, site of the present-day conflict between Ethiopia and Somalia, covers the left half of the third photograph. Small cumulus clouds form in the cool coastal air heated by the desert. *Atlantis 2*, 61B, 26 November 1985 – 3 December 1985.

123 Clouds obscure the Canadian Rockies of British Columbia south-southwest of Calgary in the province of Alberta. Portions of the Kootenay and Elk rivers are visible through the thin clouds. *Challenger 7*, 51B, 29 April 1985 – 6 May 1985.

124 The Grand Canyon of the Colorado in its entirety. The Colorado River flows downstream from Utah past Lees Ferry, Arizona, through Marble Canyon, Grand Canyon, and Virgin Canyon to Lake Mead and Black Canyon, where it turns south, forming the border between California and Arizona near Yuma. The Kaibab Plateau, the dark forested area, is to the lower right; the Coconino Plateau is at center bottom. Lake Mead and Las Vegas, Nevada, are at top center. Cloud puffs cover part of the San Francisco Peaks and the city of Flagstaff at the bottom of the photograph. The Salton Sea is at upper left. *Challenger 6*, 41G, 4 October 1984 – 13 October 1984.

125 The Kalahari Desert of Botswana in an area known as Makgadikgadi. The parallel formations are stabilized sand dunes from the Pleistocene Age. The salt lake in the lower left corner is called Soa Pan. *Atlantis 2*, 61B, 26 November 1985 – 3 December 1985.

126 The southeast coast of Brazil, about seventy miles northeast of the city of Porto Alegre. The South Atlantic Ocean has strong, highly visible currents, which move along the coast toward the north or northeast, and which created the beautiful long white barrier beach that crosses the photograph. The old coastline is now the inland shore of interior lakes created by the movement of sand and silt, which is also visible. *Skylab 3*, SL3, 28 July 1973 – 25 September 1973.

127 The Gulf of Alaska, with the great peaks of the Saint Elias Range of Alaska, Yukon, and British Columbia. Mount Logan, Canada's highest mountain peak at 19,850 feet, is to the left of the center of the photograph. Between Saint Elias Peak and Mount Vancouver, right of center, flows the great Malaspina Glacier in a great lobe of ice shaped like a human ear. *Challenger 6*, 41G, 5 October 1984 – 13 October 1984.

128 The Altiplano of northwest Argentina, not far from the border with Chile. A recent cinder cone is at the center, with windblown sand covering its northeast flank, a clear indication of the volcanic origins of the region. In this area, 100 miles west of San Miguel de Tucuman, the Altiplano is about 12,600 feet high, with peaks rising to more than 18,000 feet. Here, in the Atacama Desert, salt lakes and salt flats are common. The color of the deposited material is a strong indication of the mineral content of the rocks. *Challenger 8*, 61A, 30 August 1983 – 5 September 1983.

129 The North Sea, looking southwest toward Brittany, visible in the upper right corner. The islands of the mouths of the Rhine River are in the foreground, and above them sunlight reflects off the long causeways and dikes of the Ijsselmeer in the Netherlands. To the right is the Strait of Dover (Pas de Calais) with the southeast corner of England just above the right center edge. *Challenger 9*, 61A, 30 October 1985 – 6 November 1985.

130 The Rocky Mountains of northern British Columbia. This area of the Ruby and Thudaka ranges and the Omineca Mountains is located about 300 miles north-northwest of Prince George and about 250 miles northwest of Dawson Creek. The Fox River Canyon crosses the lower right corner of the photograph. The snowline in this December photo is at about 3,500 feet. *Columbia 6*, STS9, 28 November 1983 – 8 December 1983.

131 The high peaks of the snow-covered Andes of Chile and Argentina. Chile's capital city of Santiago is at the upper center portion of the photograph. The port city of Valparaiso hugs the Pacific Ocean. Just to the left of center is Aconcagua, a volcano at 22,834 feet, the highest peak in the Western Hemisphere. Mendoza is near the upper left center edge. *Challenger 6*, 41G, 5 October 1984 – 13 October 1984.

132 The Nubian Desert of northern Sudan and southern Egypt. This view to the northeast, taken from a point 200 miles west of Khartoum, shows the Nile from Abu Hamed downstream beyond Lake Nasser and the Aswan High Dam to the bend at Luxor. *Challenger 9*, 61A, 30 October 1985 – 6 November 1985.

Chrétien, Jean-Loup*
France
Soyuz T-6, June 1982

Cleave, Mary L.
USA
Atlantis 2, November 1985

Coats, Michael L.
USA
Discovery 1, August 1984

Collins, Michael
USA
Gemini 10, July 1966
Apollo 11, July 1969

Conrad, Charles, Jr. "Pete"
USA
Gemini 5, August 1965
Gemini 11, September 1966
Apollo 12, November 1969
Skylab 2, May 1973

Cooper, L. Gordon, Jr.
USA
Mercury 9, May 1963
Gemini 5, August 1965

Covey, Richard O.
USA
Discovery 6, August 1985

Creighton, John O.
USA
Discovery 5, June 1985

Crippen, Robert L.
USA
Columbia 1, April 1981
Challenger 2, June 1983
Challenger 5, April 1984
Challenger 6, October 1984

Cunningham, R. Walter*
USA
Apollo 7, October 1968

Demin, Lev
USSR
Soyuz 15, August 1974

Duke, Charles M., Jr.
USA
Apollo 16, April 1972

Dobrovolski, Georgi T.
USSR
Soyuz 11, June 1971

Dunbar, Bonnie J.
USA
Challenger 9, October 1985

Dzhanibekov, Vladimir A.
USSR
Soyuz 27, January 1978
Soyuz 39, March 1981
Soyuz T-6, June 1982
Soyuz T-12, July 1984
Soyuz T-13, June 1985

Eisele, Donn F.*
USA
Apollo 7, October 1968

England, Anthony W.
USA
Challenger 8, July 1985

Engle, Joe H.
USA
Columbia 2, November 1981
Discovery 6, August 1985

Evans, Ronald E.
USA
Apollo 17, December 1972

Fabian, John M.
USA
Challenger 2, June 1983
Discovery 5, June 1985

Faris, Muhammad Ahmad
Syria
Soyuz TM-3, July 1987

Farkas, Bertalan*
Hungary
Soyuz 36, May 1980

Feoktistov, Konstantin P.*
USSR
Voskhod 1, October 1964

Filipchenko, Anatoli
USSR
Soyuz 7, October 1968
Soyuz 16, December 1974

Fisher, Anna L.
USA
Discovery 2, November 1984

Fisher, William F.
USA
Discovery 6, August 1985

Fullerton, C. Gordon
USA
Columbia 3, March 1982
Challenger 8, July 1985

Furrer, Reinhard
Federal Republic of Germany
Challenger 9, October 1985

Gagarin, Yuri A.
USSR
Vostok 1, April 1961

Gardner, Dale A.
USA
Challenger 3, August 1983
Discovery 2, November 1984

Garn, Jake
USA
Discovery 4, April 1985

Garneau, Marc
Canada
Challenger 6, October 1984

Garriott, Owen K.
USA
Skylab 3, July 1973
Columbia 6, November 1983

Gibson, Edward G.
USA
Skylab 4, November 1973

Gibson, Robert L. "Hoot"
USA
Challenger 4, February 1984
Columbia 7, January 1986

Glazkov, Yuri N.
USSR
Soyuz 24, February 1977

Glenn, John H., Jr.
USA
Mercury 6, February 1962

Gorbatko, Viktor
USSR
Soyuz 7, October 1969
Soyuz 24, February 1977
Soyuz 37, July 1980

Gordon, Richard F., Jr.
USA
Gemini 11, September 1966
Apollo 12, November 1969

Grabe, Ronald J.
USA
Atlantis 1, October 1985

Grechko, Georgi G.*
USSR
Soyuz 17, January 1975
Soyuz 26, December 1977
Soyuz T-14, September 1985

Gregory, Frederick D.
USA
Challenger 7, April 1985

Griggs, S. David
USA
Discovery 4, April 1985

Grissom, Virgil I. "Gus"
USA
Mercury 4, July 1961
Gemini 3, March 1965

Gubarev, Vladimir A.
USSR
Soyuz 17, January 1975
Soyuz 28, March 1978

Gurragcha, Zhugderdemidiyn*
Mongolia
Soyuz 39, March 1981

Haise, Fred W., Jr.
USA
Apollo 13, April 1970

Hart, Terry J.
USA
Challenger 5, April 1984

Hartsfield, Henry W. "Hank," Jr.
USA
Columbia 4, June 1982
Discovery 1, August 1984
Challenger 9, October 1985

Hauck, Frederick H. "Rick"
USA
Challenger 2, June 1983
Discovery 2, November 1984

Hawley, Steven A.
USA
Discovery 1, August 1984
Columbia 7, January 1986

Henize, Karl G.*
USA
Challenger 8, July 1985

Hermaszewski, Miroslav*
Poland
Soyuz 30, June 1978

Hilmers, David C.
USA
Atlantis 1, October 1985

Hoffman, Jeffrey A.
USA
Discovery 4, April 1985

Irwin, James B.*
USA
Apollo 15, July 1971

Ivanov, Georgi*
Bulgaria
Soyuz 33, April 1979

Ivanchenkov, Aleksandr S.
USSR
Soyuz 29, June 1978
Soyuz T-6, June 1982

Jähn, Sigmund*
German Democratic Republic
Soyuz 31, August 1978

Kerwin, Joseph P.
USA
Skylab 2, May 1973

Khrunov, Yevgeni
USSR
Soyuz 5, January 1969

Kizim, Leonid
USSR
Soyuz T-3, November 1980
Soyuz T-10, February 1984
Soyuz T-15, March 1986

Klimuk, Pyotr I.
USSR
Soyuz 13, December 1973
Soyuz 18, May 1975
Soyuz 30, June 1978

Komarov, V. M.
USSR
Voskhod 1, October 1964
Soyuz 1, April 1967

Kovalyonok, Vladimir V.
USSR
Soyuz 25, October 1977
Soyuz 29, June 1978
Soyuz T-4, March 1981

Kubasov, Valeri *
USSR
Soyuz 6, October 1968
Soyuz 19-Epas, July 1975
Soyuz 36, May 1980

Laveykin, Aleksandr I.
USSR
Soyuz TM-2, February 1987

Lazarev, V. G.
USSR
Soyuz 12, September 1973
Soyuz 18-1, April 1975

Lebedev, Valentin V.
USSR
Soyuz 13, December 1973
Soyuz T-5, May 1982

Leestma, David C.
USA
Challenger 6, October 1984

Lenoir, William B.
USA
Columbia 5, November 1982

Leonov, Aleksei A. *
USSR
Voskhod 2, March 1965
Soyuz 19-Epas, July 1975

Lichtenberg, Byron *
USA
Columbia 6, November 1983

Lind, Donald L.
USA
Challenger 7, April 1985

Lounge, Michael
USA
Discovery 6, August 1985

Lousma, Jack R.
USA
Skylab 3, July 1973
Columbia 3, March 1982

Lovell, James A., Jr.
USA
Gemini 7, December 1965
Gemini 12, November 1966
Apollo 8, December 1968
Apollo 13, April 1970

Lucid, Shannon W.
USA
Discovery 5, June 1985

Lyakhov, Vladimir A. *
USSR
Soyuz 32, February 1979
Soyuz T-9, June 1983

Makarov, Oleg G. *
USSR
Soyuz 12, September 1973
Soyuz 18-1, April 1975
Soyuz 27, January 1978
Soyuz T-3, November 1980

Malyshev, Yuri V.
USSR
Soyuz T-2, June 1980
Soyuz T-11, April 1984

Mattingly, Ken
USA
Apollo 16, April 1972
Columbia 4, June 1982
Discovery 3, January 1985

McBride, Jon A.
USA
Challenger 6, October 1984

McCandless, Bruce
USA
Challenger 4, February 1984

McDivitt, James A.
USA
Gemini 4, June 1965
Apollo 9, March 1969

McNair, Ronald E.
USA
Challenger 4, February 1984
Challenger 10, January 1986

Messerschmid, Ernst *
Federal Republic of Germany
Challenger 9, October 1985

Mitchell, Edgar D. *
USA
Apollo 14, January 1971

Mullane, Michael
USA
Discovery 1, August 1984

Musgrave, F. Story
USA
Challenger 1, April 1983
Challenger 8, July 1985

Nagel, Steven R.
USA
Discovery 5, June 1985
Challenger 9, October 1985

Nelson, Bill
USA
Columbia 7, January 1986

Nelson, George D. "Pinky"
USA
Challenger 5, April 1984
Columbia 7, January 1986

Neri-Vela, Rodolfo *
Mexico
Atlantis 2, November 1985

Nikolayev, Andrian G.
USSR
Vostok 3, August 1962
Soyuz 9, June 1970

O'Connor, Bryan D.
USA
Atlantis 2, November 1985

Ockels, Wubbo J. *
Netherlands
Columbia 7, October 1985

Onizuka, Ellison S.
USA
Discovery 3, January 1985
Challenger 10, January 1986

Overmyer, Robert F. *
USA
Columbia 5, November 1982
Challenger 7, April 1985

Pailes, William A.
USA
Atlantis 1, October 1985

Parker, Robert A.
USA
Columbia 6, November 1983

Patsayev, Viktor I.
USSR
Soyuz 11, June 1971

Payton, Gary
USA
Discovery 3, January 1985

Peterson, Donald H.
USA
Challenger 1, April 1983

Pham, Tuan *
Vietnam
Soyuz 37, July 1980

Pogue, William R.
USA
Skylab 4, November 1973

Popov, Leonid *
USSR
Soyuz 35, April 1980
Soyuz 40, May 1981
Soyuz T-7, August 1982

Popovich, Pavel R.
USSR
Vostok 4, August 1962
Soyuz 14, July 1974

Prunariu, Dumitru *
Romania
Soyuz 40, May 1981

Remek, Vladimir *
Czechoslovakia
Soyuz 28, March 1978

Resnik, Judith A.
USA
Discovery 1, August 1984
Challenger 10, January 1986

Ride, Sally K.
USA
Challenger 2, June 1983
Challenger 6, October 1984

Romanenko, Yuri V.
USSR
Soyuz 26, December 1977
Soyuz 38, September 1980
Soyuz TM-2, February 1987

Roosa, Stuart A.
USA
Apollo 14, January 1971

Ross, Jerry L.
USA
Atlantis 2, November 1985

Rozhdestvenski, V. I.
USSR
Soyuz 23, October 1976

Rukavishnikov, Nikolai I. *
USSR
Soyuz 10, April 1971
Soyuz 16, December 1974
Soyuz 33, April 1979

Ryumin, Valeri V.
USSR
Soyuz 25, October 1977
Soyuz 32, February 1979
Soyuz 35, April 1980

Sarafanov, G. V.
USSR
Soyuz 15, August 1974

Savinykh, Viktor P.
USSR
Soyuz T-4, March 1981
Soyuz T-13, June 1985

Savitskaya, Svetlana E.
USSR
Soyuz T-7, August 1982
Soyuz T-12, July 1984

Schirra, Walter M., Jr.
USA
Mercury 8, October 1962
Gemini 6A, December 1965
Apollo 7, October 1968

Schmitt, Harrison H. "Jack"
USA
Apollo 17, December 1972

Schweickart, Russell L. "Rusty" *
USA
Apollo 9, March 1969

Scobee, Francis R. "Dick"
USA
Challenger 5, April 1984
Challenger 10, January 1986

Scott, David R.
USA
Gemini 8, March 1966
Apollo 9, March 1969
Apollo 15, July 1971

Scully-Power, Paul D.
USA
Challenger 6, October 1984

Seddon, Margaret Rhea
USA
Discovery 4 April 1985

Serebrov, Aleksandr A.
USSR
Soyuz T-7, August 1982
Soyuz T-8, April 1983

Sevastyanov, Vitali
USSR
Soyuz 9, June 1970
Soyuz 18, May 1975

Sharma, Rakesh
India
Soyuz T-11, April 1984

Shatalov, Vladimir
USSR
Soyuz 4, January 1969
Soyuz 8, October 1969
Soyuz 10, April 1971

Shaw, Brewster H., Jr.
USA
Columbia 6, November 1983
Atlantis 2, November 1985

Shepard, Alan B., Jr.
USA
Mercury 3, May 1961
Apollo 14, January 1971

Shonin, Georgi
USSR
Soyuz 6, October 1969

Shriver, Loren J.
USA
Discovery 3, January 1985

Slayton, Donald K.
USA
Apollo-Soyuz , July 1975

Solovyov, Vladimir A. *
USSR
Soyuz T-10, February 1984
Soyuz T-15, March 1986

Spring, Sherwood C.
USA
Atlantis 2, November 1985

Stafford, Thomas P.
USA
Gemini 6A, December 1965
Gemini 9A, June 1966
Apollo 10, May 1969
Apollo-Soyuz, July 1975

Stewart, Robert L.
USA
Challenger 4, February 1984
Atlantis 1, October 1985

Strekalov, Guennadi M.
USSR
Soyuz T-3, November 1980
Soyuz T-8, April 1983
Soyuz T-10-1, April 1983
Soyuz T-11, April 1984

Sullivan, Kathryn D.
USA
Challenger 6, October 1984

Swigert, John L., Jr.
USA
Apollo 13, April 1970

Tamayo, Arnaldo Mendez *
Cuba
Soyuz 38, September 1980

Tereshkova, Valentina V.
USSR
Vostok 6, June 1963

Thagard, Norman E.
USA
Challenger 2, June 1983
Challenger 7, April 1985

Thornton, William E.
USA
Challenger 3, August 1983
Challenger 7, April 1985

Titov, Gherman S.
USSR
Vostok 2, August 1961

Titov, Vladimir G.
USSR
Soyuz T-8, April 1983
Soyuz T-10-1, September 1983

Truly, Richard H.
USA
Columbia 2, November 1981
Challenger 3, August 1983

Van den Berg, Lodewijk
USA
Challenger 7, April 1985

Van Hofton, James D.
USA
Challenger 5, April 1984
Discovery 6, August 1985

Vasyutin, Vladimir V.
USSR
Soyuz T-14, September 1985

Viktorenko, Aleksandr S.
USSR
Soyuz TM-3, July 1987

Volk, Igor P.
USSR
Soyuz T-12, July 1984

Volkov, Aleksandr A.
USSR
Soyuz T-14, September 1985

Volkov, Vladislav
USSR
Soyuz 7, October 1969
Soyuz 11, June 1971

Volynov, Boris
USSR
Soyuz 5, January 1969
Soyuz 21, July 1976

Walker, Charles D.
USA
Discovery 1, August 1984
Discovery 4, April 1985
Atlantis 2, November 1985

Walker, David M.
USA
Discovery 2, November 1984

Wang, Taylor G. *
USA
Challenger 7, April 1985

Weitz, Paul J.
USA
Skylab 2, May 1973
Challenger 1, April 1983

White, Edward H.
USA
Gemini 4, June 1965

Williams, Donald E.
USA
Discovery 4, April 1985

Worden, Alfred M.
USA
Apollo 15, July 1971

Yegorov, Boris B.
USSR
Voskhod 1, October 1964

Yeliseyev, Aleksei *
USSR
Soyuz 5, January 1969
Soyuz 8, October 1969
Soyuz 10, April 1971

Young, John W.
USA
Gemini 3, March 1965
Gemini 10, July 1966
Apollo 10, May 1969
Apollo 16, April 1972
Columbia 1, April 1981
Columbia 6, November 1983

Zholobov, Vitali M.
USSR
Soyuz 21, July 1976

Zudov, Vyacheslav D.
USSR
Soyuz 23, October 1976

Sources

Epigraph

"I would have wished . . ." Submitted by Reinhard Furrer.

Outward

"Vitalik caresses me . . ." Thomas Levenson, "The Heart Remains on Earth," *Discover*, February 1985, Copyright © 1987 by Family Media.

"We walked across . . ." Submitted by Aleksandr Volkov.

"Q: The night before . . ." By permission of Micky Remann. Exchange between Birgit Remann and Dumitru Prunariu, Budapest, Hungary, Fall 1986.

"As we drove out . . ." By permission of Al Reinert. Interview with James Irwin, Colorado Springs, Colorado, 1982.

"They started to put . . ." Submitted by John H. Glenn, Jr.

"Then they closed . . ." By permission of Al Reinert. Interview with James Irwin, Colorado Springs, Colorado, 1982.

"The *Atlas* is . . ." Submitted by John H. Glenn, Jr.

"I heard the word ignition . . ." By permission of Al Reinert. Interview with James Irwin, Colorado Springs, Colorado, 1982.

"It seems I am leaving . . ." Valeri Ryumin, "Half a Year Away from Earth," *Soviet Life*, April 1981, p. 31.

"As soon as you get . . ." By permission of Al Reinert. Interview with Ken Mattingly, Houston, Texas, 1979.

"The rocket went farther . . ." Submitted by Zhugderdemidiyn Gurragcha.

"When the engine shut down . . ." Joseph Allen, "Joe's Odyssey," *Omni*, June 1983, p. 114. Copyright © 1983 by Joseph Allen and reprinted with the permission of Omni Publications International Ltd.

"Suddenly I saw . . ." Submitted by Jeffery Hoffman, from his book, *An Astronaut's Diary*, Montclair, N.J.: Caliban Press, 1985.

"Weightlessness comes on abruptly . . ." Submitted by Miroslav Hermaszewski.

"I used to have dreams . . ." Courtesy of "Frontline," from their television program, "The Real Stuff," January 27, 1987.

"We orbit and float . . ." Submitted by Joseph Allen, from his article, "Taking the High Road," *Air and Space Magazine*, February/March 1987, pp. 8, 9.

"I enjoyed the fact . . ." By permission of Al Reinert. Interview with Michael Collins, Washington, D.C., 1981.

"On the way back . . ." By permission of Al Reinert. Interview with Charles Duke, New Braunfels, Texas, 1982.

"We spent most of . . ." From *Apollo 10* Crew Press Conference, June 7, 1969, Apollo News Center, Houston, Texas.

Observations

"It seems to me . . ." Submitted by Pavel Popovich.

"Several days after . . ." Submitted by Igor Volk.

"For the first time . . ." Submitted by Ulf Merbold.

"You see layers . . ." Joseph Allen, "Joe's Odyssey," *Omni*, June 1983, p. 63. Copyright © 1983 by Joseph Allen and reprinted with the permission of Omni Publications International Ltd.

"They say if you . . ." Submitted by Taylor Wang.

"You realize that . . ." Submitted by Yuri Romanenko, received from orbit.

"Flying above the planet . . ." Valeri Ryumin, *175 Days in Space: A Russian Cosmonaut's Private Diary – And an Incredible Human Document*, pp. 10, 11. By permission of Mr. Henry Gris.

"This beauty consists . . ." Submitted by Patrick Baudry.

"The Pacific . . ." U.S. and Canada editions: from *A House in Space* by Henry S. F. Cooper, Jr. Copyright © 1976 by Henry S. F. Cooper, Jr. Reprinted by permission of Henry Holt and Company, Inc. Foreign editions excluding the British Commonwealth: Copyright © 1976 by Henry S. F. Cooper, Jr. From *A House in Space*, published by Holt, Rinehart and Winston. British Commonwealth editions: from *A House in Space* by Henry S. F. Cooper, Jr. Reprinted by permission of A. M. Heath and Company, Ltd.

"Although the ocean's surface . . ." English editions: James E. Oberg and Alcestis R. Oberg, *Pioneering Space*, New York: McGraw-Hill, 1986. Foreign editions: James E. Oberg and Alcestis R. Oberg, *Pioneering Space*, c/o Richard Curtis Associates, New York.

"Now as we were crossing . . ." U.S. and Canada editions: from *A House in Space* by Henry S. F. Cooper, Jr. Copyright © 1976 by Henry S. F. Cooper, Jr. Reprinted by permission of Henry Holt and Company, Inc. Foreign editions excluding the British Commonwealth: Copyright © 1976 by Henry S. F. Cooper, Jr. From *A House in Space*, published by Holt, Rinehart and Winston. British Commonwealth editions: from *A House in Space* by Henry S. F. Cooper, Jr. Reprinted by permission of A. M. Heath and Company, Ltd.

"We were able to see . . ." U.S. and Canada editions: from *A House in Space* by Henry S. F. Cooper, Jr. Copyright © 1976 by Henry S. F. Cooper, Jr. Reprinted by permission of Henry Holt and Company, Inc. Foreign editions excluding the British Commonwealth: Copyright © 1976 by Henry S. F. Cooper, Jr. From *A House in Space*, published by Holt, Rinehart and Winston. British Commonwealth editions: from *A House in Space* by Henry S. F. Cooper, Jr. Reprinted by permission of A. M. Heath and Company, Ltd.

"In the middle of the night . . ." Submitted by Robert Cenker.

"As we were flying . . ." Submitted by Lev Demin.

"A cosmonaut . . ." Submitted by Dumitru Prunariu.

"It isn't important . . ." Submitted by Yuri Artyukhin.

"When the Russian cosmonaut . . ." By permission of Micky Remann. Interview with Ernst Messerschmid, Budapest, Hungary, Fall 1986.

"The sunlight on . . ." Submitted by Karl Henize.

"The signs of life . . ." Submitted by Marc Garneau.

"Madagascar is still . . ." Submitted by Karl Henize.

"Africa looked ill . . ." By permission of Micky Remann. Interview with Robert Overmyer, Budapest, Hungary, Fall, 1986.

"Dasht-i-Kavir . . ." Robert L. Crippen and John W. Young, "Our Phenomenal First Flight," *National Geographic*, October 1981, p. 494.

"After an orange cloud . . ." Submitted by Vladimir Kovalyonok.

"From space I saw Earth . . ." Submitted by Muhammad Ahmad Faris.

"We are passing over . . ." Submitted by Valentin Lebedev.

"The clouds were always . . ." U.S. and Canada editions: From *A House in Space* by Henry S. F. Cooper, Jr. Copyright © 1976 by Henry S. F. Cooper, Jr. Reprinted by permission of Henry Holt and Company, Inc. Foreign editions excluding the British Commonwealth: Copyright © 1976 by Henry S. F. Cooper, Jr. From *A House in Space*, published by Holt, Rinehart and Winston. British Commonwealth editions: From *A House in Space* by Henry S. F. Cooper, Jr. Reprinted by permission of A.M. Heath and Company, Ltd.

"One expects blues . . ." Submitted by Byron Lichtenberg.

"My mental boundaries . . ." Submitted by Rakesh Sharma.

"Around six pm . . ." Submitted by Jean-Loup Chrétien.

"The first day . . ." Submitted by Sultan Bin Salman al-Saud.

"From space I see . . ." Submitted by Rodolfo Neri-Vela.

"I was only twelve . . ." Submitted by Bertalan Farkas.

"When you look out . . ." Submitted by Loren Acton.

"After eighteen days . . ." Submitted by Yuri Glazkov.

"I have been in love . . ." Submitted by Pham Tuan .

"As I looked down . . ." Submitted by John-David Bartoe.

"During a space flight . . ." Submitted by Boris Volynov.

Space Stations

"We approached the station . . ." Submitted by Georgi Ivanov.

"I had to accomplish . . ." Submitted by Viktor Gorbatko.

"O. Henry . . ." Valeri Ryumin, *Six Months Above the Planet*, p. 1. By permission of Mr. Henry Gris. "Entering *Salyut* . . ." Valeri Ryumin, "Half a Year Away from Earth," *Soviet Life*, April 1981, p. 31.

"The first hours in space . . ." Submitted by Georgi Beregovoy.

"It is amusing . . ." Submitted by Andriyan Nikolayev.

"After the third . . ." Submitted by Viktor Savinykh.

"We left the spacecraft . . ." Submitted by Vladimir Dzhanibekov.

"The station had fallen . . ." Submitted by Vladimir Dzhanibekov.

"For seven long days . . ." Submitted by Viktor Savinykh.

"Not a day without . . ." Submitted by Vladimir Kovalyonok.

"During my flight . . ." Submitted by Vladimir Lyakhov.

"Today we welcomed . . ." Valeri Ryumin, *175 Days in Space: A Russian Cosmonaut's Private Diary – And an Incredible Human Document*, p. 5. By permission of Mr. Henry Gris.

"We found dried apricots . . ." English editions: James E. Oberg and Alcestis R. Oberg, *Pioneering Space*, New York: McGraw-Hill, 1986, p. 248. Foreign editions: James E. Oberg and Alcestis R. Oberg, *Pioneering Space*, c/o Richard Curtis Associates, New York.

"In order to show . . ." Submitted by Valeri Kubasov.

"We had various kinds . . ." V. Gor'koy and N. Kon'kov, "Cosmonaut Berezovoy's Memoirs on 211-Day Spaceflight," *Aviatsiya I Kosmonavtika* 7 (July 1983), 8 (August 1983), 9 (September 1983).

"In the morning . . ." Submitted by Vitali Sevastyanov.

"And how's the weather . . ." V. Gubarev, "Five Months into the Flight: Berezovoy and Lebedev Comment on Five Months in Space," *Pravda*, October 13, 1982, p. 3.

"All day long . . ." By permission of Micky Remann. Interview with Oleg Makarov, Budapest, Hungary, Fall 1986.

"The *Salyut* station . . ." Submitted by Yuri Artyukhin.

"I once saw ice crystals . . ." Submitted by Vitali Sevastyanov.

"Lebedev had never . . ." V. Gor'koy and N. Kon'kov, "Cosmonaut Berezovoy's Memoirs on 211-Day Spaceflight," *Aviatsiya I Kosmonavtika* 7 (July 1983), 8 (August 1983), 9 (September 1983).

"The first shoots . . ." Thomas Levenson, "The Heart Remains on Earth," *Discover*, February 1985. Copyright © 1987 by Family Media.

"It's impossible to imagine . . ." Submitted by Georgi Grechko.

"We brought some small fish . . ." Submitted by Vitali Zholobov.

"From orbit we observed . . ." V. Gor'koy and N. Kon'kov, "Cosmonaut Berezovoy's Memoirs on 211-Day Spaceflight," *Aviatsiya I Kosmonavtika* 7 (July 1983), 8 (August 1983), 9 (September 1983).

"One morning I woke up . . ." Submitted by Aleksandr Aleksandrov.

"After several weeks . . ." Submitted by Pyotr Klimuk.

"While out there . . ." English editions: James E. Oberg and Alcestis R. Oberg, *Pioneering Space*, New York: McGraw-Hill, 1986, p. 250. Foreign editions: James E. Oberg and Alcestis R. Oberg, *Pioneering Space*, c/o Richard Curtis Associates, New York.

"Above my sleeping bag . . ." Submitted by Leonid Kizim.

"When I had a free minute . . ." Submitted by Anatoli Berezovoy.

"In order to see . . ." Submitted by Arnaldo Tamayo Mendez.

"You talk with . . ." V. Zubkov, "I and My Comrade: Post-Flight Interview with Berezovoy and Lebedev," *Sotsialisticheskaya Industriya* 26 (December 1982), p. 4.

"In the midst of . . ." Submitted by Valeri Kubasov.

"Once during the mission . . ." Submitted by Vitali Sevastyanov.

"Today we had . . ." Thomas Levenson, "The Heart Remains on Earth," *Discover*, February 1985. Copyright © 1987 by Family Media.

"On the floor . . ." Submitted by Oleg At'kov.

"Night, and we had . . ." Submitted by Aleksandr Laveykin.

"Our mission was coming to . . ." Submitted by Svetlana Savitskaya.

"The moments of parting . . ." Submitted by Aleksandr Viktorenko.

"A year has passed . . ." Valeri Ryumin, *175 Days in Space: A Russian Cosmonaut's Private Diary – And an Incredible Human Document*, p. 1. By permission of Mr. Henry Gris.

"We have entered . . ." V. Gor'koy and N. Kon'kov, "Cosmonaut Berezovoy's Memoirs on 211-Day Spaceflight," *Aviatsiya I Kosmonavtika* 7 (July 1983), 8 (August 1983), 9 (September 1983).

"As we encounter . . ." Submitted by Gordon Fullerton.

"When the spacecraft . . ." Submitted by Lev Demin.

"A we descended farther . . ." Valeri Ryumin, *175 Days in Space: A Russian Cosmonaut's Private Diary – And an Incredible Human Document*, p. 18. By permission of Mr. Henry Gris.

"Space is so close . . ." Submitted by Wubbo Ockels.

"A strange feeling . . ." Submitted by Andriyan Nikolayev.

"Once again I stand . . ." Submitted by Georgi Shonin.

Reflections

"When I came back . . ." By permission of Al Reinert. Interview with James Irwin, Colorado Springs, Colorado, 1982.

"The first two nights . . ." Submitted by Thomas Stafford.

"The peaks were . . ." By permission of Al Reinert. Interview with Edgar Mitchell, West Palm Beach, Florida, 1981.

"As the journey . . ." Submitted by Vladimir Solovyov.

"When the history . . ." William Stockton and John Noble Wilford, "Space and the American Vision," *The New York Times Magazine*, April 5, 1981. Copyright © 1981 by the New York Times Company. Reprinted by permission.

"The winds scatter . . ." Submitted by Yuri Glazkov.

"In space one has . . ." Submitted by John H. Glenn, Jr.

"We went to the moon . . ." Submitted by Edgar Mitchell.

"Instead of an intellectual search . . ." By permission of Stanley Rosen. Interview with Edgar Mitchell, Palo Alto, California, July 1974. "On the return trip home . . ." Submitted by Edgar Mitchell.

"I know the stars . . ." By permission of Al Reinert. Interview with Eugene Cernan, Houston, Texas, 1981 and 1982.

"For those who have . . ." Submitted by Donald Williams.

"When we look into . . ." Submitted by Vladimir Shatalov.

"Before I flew . . ." Submitted by Sigmund Jähn.

"Of all the people . . ." Submitted by Robert Cenker.

"And then you look back . . ." Submitted by Russell Schweickart.

"And tomorrow? . . ." Submitted by Yuri Gagarin.

Acknowledgments

Since its conception, this book has been nurtured by an international family of talented, dedicated people with whom it has been a privilege to work.

My foremost thanks go to all the astronauts and cosmonauts who, despite heavy work loads, responded to our invitation to contribute. Yuri Romanenko and Aleksandr Aleksandrov, for example, were in orbit when the text was being prepared, yet they took time out from a long and arduous mission to offer their observations.

From the beginning, Alan Gump, Paul and Barbara Kayfetz, Michael Lerner, Lisa O'Malley, Joseph Miller, Don and Martha Rosenthal, and Marion and Warren Weber have provided encouragement and support. Alan and Don in particular were my faithful guides in the area of concept development.

Special acknowledgment must be made to the Institute of Noetic Sciences in Sausalito, California, whose sponsorship was crucial to the project's success. An organization with world-wide membership dedicated to understanding the nature of consciousness and the relationship of body to mind – "inner space" – the Institute was founded by Astronaut Edgar Mitchell, a co-founder of the Association of Space Explorers. Committed to the study of the nature of healing and of health, and to exploring the farthest reaches of human nature, Noetic Sciences works toward a positive global future. Thanks to Executive Director Wink Franklin and the Board of Directors for their assistance and commitment in the creation of this work.

Oleg Makarov and Byron Lichtenberg of the Association of Space Explorers Book Committee deserve thanks for their faith, enthusiasm, and steadfast hard work, which ensured the success of our efforts. Rusty Schweickart added practicality to generosity in loaning me his personal lap-top computer, which enabled me to communicate on-line with the astronauts and cosmonauts. And my deepest gratitude to Loren Acton, whose enthusiasm, vision, and wise counsel guided us with success around many an obstacle. Thanks, too, to Evelyn Acton for her good cheer and patience, and to ASE's communication consultant Alan Kelly. For his quick wit and good-humored help, and for bearing much of the burden of our work, I'm most grateful to the Association's Administrative Director, Ted Everts.

In the course of this project, I have been blessed to meet special individuals with whom I've shared a communion of vision and purpose. That communion, which has shaped this book, has shaped me as well. I feel privileged to have worked with Lt. Col. Stanley Rosen, Director of Long-Range Planning, Air Force Space Division, and filmmaker Al Reinert, who were as magnanimous with information as they were with enthusiasm and encouragement. The life and work of Captain Jacques-Yves Cousteau has long been an inspiration. I am most grateful for his invaluable observations on the early manuscript and for his participation in this book.

I'm indebted, too, to B. J. Bluth and Martha Helppie for their generosity with the collected experiences of cosmonauts in space, to James Oberg for sharing his vast store of information, and to Henry Gris for his efforts on the book's behalf. For counsel that kept the project perking, thanks to Roberta Cairney and Christopher Thorsness.

And for their unsparing personal support, my sincerest thanks to Michael Broffman and Jay Rice.

Dick Underwood, perhaps the world's most knowledgeable guide to the vast and complex world of earth photography, has been a perpetual source of help and inspiration to me. I must also extend my gratitude to all those individuals at the NASA Johnson Space Center in Houston for their professionalism and fellowship, and especially to Doug Ward and Bill Robbins, Public Affairs, and John Holland, Technology Division, who provided me access to NASA's exhaustive photo archive. For their assistance in helping me find the best NASA images – the extraordinary photos you see here – my deepest appreciation to Mike Gentry, Lisa Vasquez, Eileen Walsh, Cynthia Leos, and Carolyn Fields of Media Services, and to Barry Schroder of Technicolor Graphic Services, as well as Charles Shrimplin, Richard Tousignant, Terry Slezak, Barney Corbin, and Johnny Salinas, master technicians and imagists all. And special thanks to Jon Schneeburger, illustrations editor at *National Geographic* magazine, and Rose Aiello, photo librarian at the Center for Earth and Planetary Studies at the Air and Space Museum, whose last-minute help in acquiring photos saved the day.

I owe much to the team of dedicated professionals who handled the nuts and bolts of the book's preparation – my deepest gratitude to Bruce Dearborn, Connie Hoshor, and Candace Wyatt, my faithful office staff and mainstays of motivating spirit and good humor. They bore the burden at all hours, coordinating a mind-boggling complex of details, and they always came through on time and on

the mark. They deserve much of the credit for this book.

Much credit is also due to Debbie Morrison, Joan Bertsch, and Lila Phillips of Fineline Office Services, as well as to Mary Bruce, Susan Goldhaber, Patrice Daley, Gail Moss, Nancy Rhine, Meg Simmons, Joel Kramer, Tim Gates and Denise Feldman, Jim Anderson, Ken and Sarah Masterton, Bird Brother, Jenny Ulrich, Beth Clymer, Jenny Cliner, Doug Perrin, Lara Braun, and Faulkner's Color Lab.

My sincere gratitude to the publishing firm of Addison-Wesley goes without saying, but I'd like to name a few members of the company to whom I'm especially grateful: Executive Vice-President Ann Dilworth, Publisher David Miller, Director of Marketing George Gibson, Carolyn Savarese in Subsidiary Rights, Robert Shepard in Special Markets, Lori Foley in Production, Diane Hovenesian in Publicity, and Editorial Assistant John Bell.

Stuart Miller, Addison-Wesley's west coast acquisitions editor, deserves thanks as the perfect matchmaker for bringing Addison-Wesley and this project together, and Editorial Director Jane Isay deserves thanks for introducing me to the exigencies of publishing. I'm deeply grateful to William Patrick, my editor at Addison-Wesley, whose superlative talents as a writer assured the text's precision and grace, and to Douglass Scott for the grace and elegance of his design.

Others related to the book industry who offered wise counsel, and for whom I'm most appreciative, are publishers Don McQuiston of McQuiston & Daughter, Rick Smullen and David Cohen of Collins Publishing Co., Andy Stewart of Stewart, Tabori and Chang, printbroker Ken Coburn of Interprint, book designer James Stockton, and publishing consultant Martin Levin, whose fatherly advice inspired me to new levels of self-discovery. Publisher Jon Beckmann of the Sierra Club was especially helpful and generous with his talent and advice, as was writer Peter Carroll and editors Micky Remann and John Robbins, all of whom deserve my very special thanks. Thanks for additional editorial assistance from Charles Fox and Doris Ober.

Carol Denison deserves special mention and thanks for her creative advice and encouragement throughout the project.

I'm perhaps most grateful to Brad Bunnin, who has been more like the book's guardian than its attorney and negotiator, and whose sagacity and diplomacy make him a trusted counselor in the truest sense of the word.

Among the international family who provided logistical support for this book were Cosmonaut Sigmund Jähn, Astronaut Wubbo Ockels and the French Space Agency, and Dr. Roger Bonnet, Director of the Scientific Program at the European Space Agency.

Because the subject of the book knows no political borders or boundaries, I was able to venture across several in its preparation. Senator Claiborne Pell and Dr. Scott Jones were extremely helpful to me in that regard, along with Consul General Viktor Kamenev and Vice-Consul Sergei Aivazian of the Soviet Consulate in San Francisco, and Oleg Benyukh, Chief Information Officer at the Soviet Embassy in Washington, D.C.

Other partners in communication with my Soviet colleagues were Joel and Diane Schatz of San Francisco/ Moscow Teleport, who provided a direct telecommunications link between Mir, Addison-Wesley, and myself; Henry Dakin, Cindy Spoor, and the other talented associates at H.S. Dakin & Co., who have my admiration for their dedication to furthering understanding between our two nations; translators Dr. Andrei A. Voronkov, Senior Research Fellow at the Institute of the USA and Canada Academy of Sciences of the USSR, and Jan Butler. I owe special thanks to Natasha Ward and Tanya Khotin for their dedication and commitment in interpreting and translating, as well as to Anya Kucherova with US Information Moscow and Tom Jenkins of Newgate Travel.

Time-Life Soviet Bureau is to be credited for their kind assistance as couriers of material between Mir, Addison-Wesley, and myself.

In the Soviet Union, I was the recipient of the good will and hospitality of some extraordinary people. I'm very thankful to Cosmonaut Aleksei Leonov for introducing our delegation so graciously to Star City; and to Mikhail Nenashev, Chairman of the USSR State Committee for Printing, Publishing and Book Trade, and the Committee's First Deputy Chairman, Dmitri Mamleyev. I'm most appreciative of Public Relations Director Tankrid Golonpolsky, the consummate diplomat, accommodating always, and extremely helpful to our project.

My sincere thanks also to our hosts Andrei Kokoshin, Deputy Director of the Institute of USA/Canadian Studies at the prestigious USSR Academy of Sciences, Yuri A. Ushanov, Ph.D., and Sergei Pouzanov, Senior Researchers at the Institute; and to Andrei A. Rexinin of the Committee for Publishing's Protocol Department.

I owe a similar debt of thanks to Mir Publishers in the Soviet Union as I do to Addison-Wesley in the States, and special appreciation for Mir's Director, Vladimir P. Kartsev, an enthusiastic supporter of this project and a most charming man. Dr. Guennady B. Kurganov, Chief Editor and consummate negotiator and facilitator, and Yuri L. Maximov, Mir's Chief of Design, also deserve special acknowledgment for their efforts on the book's behalf.

I would also like to mention the efforts of Vladimir Denisenko from the Intercosmos Council and to Vyacheslav Propoi, Dmitri Zaidin, Aleksandr Gvozdievsky, Boris Gerasimov, Aleksandr Burdakov, and Yuri Kuzmin from Mir Publishers. Coordinating the activities of so many people and keeping a complex project like this moving requires an abundance of energy and administrative skill; for this my gratitude goes to the Deputy Chief Editor of Mir Publishers, Donara Zakharova.

I had the cheerful cooperation of two agencies to help me with photo research in the USSR. At Novosti Press, General Director Valentin Falin, Photo Chief Evgueni Bogomolov, Vice-Head V. A. Verbenko, and Senior Editor B. G. Vlasov were especially helpful. Of extreme help as well at Novosti were Feydor Svetanov, Tatylana Popova, and Alex Duzofeyev, Deputy Chief of the Industrial Department. TASS was also responsive and supportive in providing photo resources, for which I'm grateful to Viktor N. Zatseppin, Editor-in-Chief of TASS's Fotokhronika. My special thanks to Mrs. Margarite Vinogradova, Director of TASS's Film Library, and to Paul Petrov at the Export Photo Desk.

The Russian and English editions of this book are identical in content, even though each contains several languages. This achievement was made

possible by the efforts of the translators. I must express my thanks to
Anatoly Weber for high-quality trans-
lations from English into Russian, and
to Richard Hainsworth, who rapidly,
enthusiastically, and creatively trans-
lated the Russian texts into English,
on occasion working through the
night to meet courier deadlines. The
translation efforts were assisted by
colleagues at Raduga Publishers,
namely Viktor Khmyzov, Vera
Selivanova, Natalia Chukanova, Lilia
Khomutova, and Tibery Borzho, and
by colleagues at Mir Publishers,
namely Galina Bikulova, Galina
Zakharova, Elena Troshneva, Svetlana
Landau, Galina Matveyeva, Devendra
Prasad Verma, Muhammad Javad Al-
muhammad, Lev Glotov, Fedor Petrov, Mark Chernyak, and Aleksandr
Yastrebov.

One of the great rewards in working
on this project has been my introduc-
tion to Elena Knorre, who did much of
the initial work on the cosmonauts'
manuscripts.

My sincere thanks go to Lev De-
sinov, Yuri Kienko, and Viktor Per-
mitinov from the Nature State Center
for their cooperation in the selection
and preparation of high-quality slides.
Anatoli Lepikhov rendered valuable
help in gathering and editing some of
the material.

My particular appreciation must
go to Yuri Voronov for his inspired
and dedicated efforts in editing the
Russian-language material.

And special thanks to my wife
Susan, who participated at all levels.
Her wise counsel, encouragement, and
abiding love sustained me through a
long and sometimes difficult project.
Special thanks belong, as well, to my
children Vanessa and Aaron, for their
participation, good advice, and love.

My father and mother helped lay the
foundations for this book by sharing
their love of nature and beauty. I thank
my father for passing on his interest in
science and for teaching me how to get
things done. I thank my mother for im-
parting to me her deep appreciation of
the mysterious and the unfathomable.

To all who have helped – you have
made me understand that there is
nothing that cannot be accomplished
where there is vision, commitment,
and dedication. This now is my greatest source of hope for all the world.

Kevin W. Kelley

THE HOME PLANET was edited by William Patrick.
Editorial assistance was provided by John Bell.

Production was supervised by Lori Foley.
Manufacturing was coordinated by Ann DeLacey.

The design was created and executed by
Douglass G. A. Scott of WGBH Design, Boston.
Design assistance was provided by Jeanne Lee and
Matthew Bartholomew.

The English text of this book was set in
Trump Mediaeval by M.J. Walsh of WGBH Design.
The foreign text was set by Spectrum
Composition Services, New York.

The book was printed and bound in Italy by
Arti Grafiche Amilcare Pizzi, S.p.A., Milan.
The acid-free text paper, 150 gsm Gardagloss,
was manufactured by Cartiere del Garda.